THE
FIELD AND GARDEN
GUIDE TO HERBS

The
Field and Garden
Guide to Herbs

M. M. Kondor
and
C. B. Wilson

Stackpole Books

Published by
STACKPOLE BOOKS
Cameron and Kelker Streets
P. O. Box 1831
Harrisburg, PA 17105

Printed in the U.S.A.

Library of Congress Cataloging in Publication Data

Kondor, M. M.
   The field and garden guide to herbs.

   Includes index.
   1. Herbs. 2. Herb gardening. 3. Herbals.
4. Herbs—North America. 5. Herb gardening—North
America. 6. Herbals—North America. I. Wilson, C. B.
II. Title.
SB351.H5K66 1983   582'.063   83-471
ISBN 0-8117-0655-9

Contents

ʃ

Foreword

Euell Gibbons proved that foraging is much more than a casual weekend hobby. The simple point is that both the birth and survival of almost every major nation originally depended upon the use and development of products from its native plants! America, for example, still uses the mature seedpods from the Common Milkweed plant to make parachutes. Rubber, which was in extremely high demand during World War II, was also extracted from the Milkweed in the raw form. Even the Common Dandelion was considered as a secondary source of rubber. During the Cold War, the chemical originally used to make the dramaticized "truth serum" was prepared from the roots of plants from the Nightshade family. America has, throughout her history, depended upon the use of native plants to supply needed raw materials and food.

Imagine a small weather-beaten troop of soldiers trying to survive the very harsh winter of 1777 in Valley Forge, Pennsylvania. Both food and clothing were in very short supply. The local immigrants, most of whom were of German descent, brought along with them from their native country the plant foods and medicines they needed, since they did not know what they would find in their newly found land. It is a well-documented fact that meals were regularly prepared and fed to the starving troops by these local residents. But through poor gardening practices, or the abandonment of gardens altogether, many of these foreign plants escaped into the wilds, only to be recognized later as the "weeds" that yearly haunt most gardens.

Plants such as Lamb's Quarters, Shepherd's Purse, and Purslane are regarded as epicurean treats in most countries where they grow. But despite the importance of these plants in our nation's growth, they are, for the most part today, considered to be simple weeds.

What, then, is a weed? Modern agriculturalists would have us believe that a weed is a plant that is growing "out of place." But this definition is both misleading and untrue. A wild plant depends on so many different environmental conditions, that, a plant growing out of place would probably die. Remember how freely wild plants grow! It is the garden you are trying to create that is the invader of a territory, forcing out the wild weeds for a period of time. In this instance, it is the garden plants that are growing out of place.

I would, instead, substitute for the above definition the statement that a weed is nothing more than a *plant without utility*. If a plant cannot be used for anything, then it must be a weed! The fact is that the place where a plant is actually growing has little or nothing to do with its being a weed. If you cannot grow a plant, or decide not to use it, or know not how to use it, then, to you the plant is a weed. But someone who may have more luck, or know more, may place a value on the plant. So, to twist an old adage, I could say that one man's weed is another man's treasure! Let's go one step further. Imagine a cornfield in which some beautiful wild roses have sprung up. Which plants are the weeds, the roses or the corn? If the roses are more useful to you for their beauty, then you have a rose garden full of very large weeds. But if a bumper crop of corn means more to you than all the roses in the world, then the roses in your cornfield are weeds! You can easily see how an individual's interpretation of a particular plant's utility can be an important consideration when attempting to classify it as a weed. In this light, I can honestly say that I have not been able to find a weed for the last ten years!

The following 200 entries are all herbs, except for the addition of a few seeds and roots. For example, Chicory can be used as an herb or as a root plant: the entire plant is thoroughly discussed. Ginseng, however, is only ever used for its root, never any other part to great extent. Ginseng, for reference purposes, is included in this complete listing of North American herbs. An "authority" who includes Ginseng in a book of herbs, however, should be sure to remind his readers of the definition of an herb. I refer not to the slang definition, but to the proper and more acceptable definition taken from the *Compact Edition of the New Oxford Dictionary*. It is actually three definitions and states that an herb is (1) a leafy part of a plant, especially distinct from the roots; (2) a plant of which the leaves and stems are used for food, medicine, scent, or flavor; and (3) a stem not woody and persistent, as in a shrub or tree, but that remains more or less succulent and dies down to the ground or entirely after flowering. While Ginseng does not fit any one of the above definitions, it is, nonetheless, included in this text because of the interest it evokes, and its extreme popularity and value. As for the rest of the entries, they make up what may be one of the most complete listings of natural herbs available to you today.

The colonists who believed in America depended on their herbs as much as their guns and armory. The immigrants, who brought many herbs with them on the journey to the new land, so brought food, clothing, and medicine for our growing nation. And perhaps the meager meals provided to our troops from herbs enabled them to survive when otherwise they could not, and thus our national spirit survived.

To cast off foraging as a simple fad, or worse yet, a casual pastime, is to cast off an idea that may have assisted in the development of entire nations. Foraging for herbs may have historical significance, be amusing, or create fun. But consider for a moment how important it could be to have some knowledge of the herbs and thus gain the independence and security that being able to survive and live on your own provides! This text will inform you, and it may entertain you, but it could also give you the edge to survive.

Thomas G. Arehart
President
Euell Gibbons Environmental Foundation

Preface

The need for a somewhat more practical and "Americanized" version of a book of herbal information became apparent to us after a survey of books on herbs. So often, the finer things in life are experienced only by those who have the knowledge or financial means to procure them. But herbs are Nature's gift to all mankind, and with a little help, they can be enjoyed by all, not just an elitist few. Recent trends indicate that many Americans desire a chance to break away from the synthetic jungle of modern products and conveniences and to delve into the natural world of wonders that lie beyond the bare concrete and asphalt boundaries of civilized life. But how? And why? A person looking for something new to do can easily be overwhelmed by the sight of myriad unfamiliar plants that seem to blanket the heretofore unknown.

Most herbals available today offer singular specialties: some are a fabulous aid to identification, others provide gobs of information, but no pictures of plants with which to associate that information, and still other provide both pictures and informative data, but on plants found only in European countries. So the task, as we saw it, was to produce a text that would expand on the books that were lacking, and make more practical, relevant, and concise the books that were not.

We decided to produce a book of herbal information that the average North American could actually use from day to day. We felt the book should provide the information that would allow anyone to actually go out there and find a particular herb, or to grow a particular herb for himself. And we wanted to supply the pertinent and useful information necessary to give that person the knowledge and understanding to assure him that he could use that herb if the need arose.

We talked to people who know many things about herbs, and, indeed, have devoted their lives to the subject; and we read of people who also know much about the subject. Admittedly, no one knows everything about herbs. At every step of the way, however, we attempted to corroborate and substantiate every bit of information available to us. Much of that information dates back to the ancient Greeks, and much of it hasn't changed from that time. Much of the information available today conflicts, and much of it is known to be erroneous, so that the task was not an easy one. This text is the result of our efforts.

We would like to acknowledge those who contributed to those efforts:

Thomas G. Arehart, B.S. Pennsylvania State University, president and founder of the Euell Gibbons Environmental Foundation, and self-taught master of aspiring herbalists, who provided the basis from which we developed this text, as well as the knowledge he has acquired from years of personal experience with herbs, and the wisdom and advice that only a master of his craft can provide.

Dr. Frederick G. Myers, and Ms. Holly Shimizu, of the United States National Arboretum, Washington, D.C., who provided much of the information on current uses of herbs, and who gave us a feeling for the "mood" of the country concerning herbs. The National Herb Garden, a project of the Herb Society of America, is found within the Arboretum complex, and provided much visual and informational stimuli.

The Stitler family, especially Kitty, who supplied assistance and photographic expertise.

Jacob Henry, who gave us the benefit of his 80-year life experience in foraging, especially for Ginseng and Golden Seal; and his daughter Betty, and the entire Henry family, who went out of their way to assist us.

Ruth Hyatt, of the Amish House and Farm in Lancaster, PA, who provided much herbal literature and gave us the benefit of years of experience with herbs.

Walter and Jesse Tyler, who allowed us to roam over their property in search of hidden herbs.

Moses Hostetler, who gave us a guided tour of the Appalachian Mountains near his Amish homestead.

Jim Bruehl, who assisted with the photography.

Michelle Toole, who made a long journey to assist in the production of the text.

Kim Watters, who provided photographic equipment.

Our families, who provided encouragement and support every step of the way.

And Judy Schnell, who gave us the opportunity to do this, and whose expert copy editing transformed a raw manuscript into a workable text.

Introduction

The material presented in this text is intended for informational purposes *only*! We are, in effect, merely reporting what is known, what is suspected, and what has long been said about the more than 200 herbs and plants listed hereafter. Herbs have long been shrouded in a blanket of mystery and superstition, and while their culinary and medicinal applications have been documented, the fact remains that much of the information available on them is contradictory and confusing. Whenever possible, we have drawn on several sources and have used only information that could be confirmed by more than one source.

We are not professing that you, the reader, abandon your fast-food restaurants in favor of a romp through the nearest field or forest; nor are we saying that you should abandon your general practitioner in favor of "natural" healing herbs found in the outdoors. Indeed, we are not saying those things at all.

What we are saying is that, here is a list of numerous common herbs and plants, and this is what our experience and the experience of others has shown to be true of them. We think this information is both interesting and entertaining, and it may offer the curious individual something new to delve into.

Many of the herbs listed herein are extremely dangerous, and many are not. Belladonna and Wolfsbane, for example, can cause violent illness and near-certain death if ingested in sufficient quantity, while Parsley and Oregano have been used for centuries to the benefit of taste buds all over the world. We think it's important to know these facts and to use caution when dealing with herbs and plants, just as you would with any unknown.

We found that many of the herbs and plants we wanted to deal with are known by several common names. In general, in those cases, we went by the most commonly used "common" name when referring to the plant, or, in toss-up situations, we went by the common name we were most familiar with. We have, however, listed alternate common names for many of the herbs and plants found within, for the benefit of readers in different parts of North America.

As you may have noticed on the preceding Contents page, the herbs and plants in the main section of the book are arranged in alphabetical order by scientific family names; and, correspondingly, the members of each family are arranged alphabetically by genus, then species name.

What *is* a family, a genus, and a species? The family is a group of plants with a closely related, recognizeable structure, a genus is a group of plants of one family with somewhat more specific common characteristics, and a species is, for our purposes, an individual member of a particular genus.

In the appendix in the back of the book you will find the herbs and plants listed alphabetically by common name, with corresponding page numbers in the text. This will probably be the most useful table for the novice who is familiar with only the common names of most plants. A glossary of botanical terms follows the appendix.

Welcome to the world of herbs!

Chapter 1

The History of Herbs

It's a safe bet that man's first experiences with herbs were the result of a continuing search for food. The sampling of potential food sources must have been quite unscientific and extremely hazardous. Perhaps early man observed what animals consumed and then decided to try those plants himself. Or, perhaps he just took a chance on eating plants that looked appetizing. Who knows? Maybe some primitive natural impulse that modern man has long since suppressed drove early man to seek or pick out a particular plant for consumption. And just as an ailing dog often consumes grass (for whatever reason), so too, perhaps, did early man instinctively seek out a particular plant for its inherent medicinal properties.

As man became more civilized, perhaps he lost some of his instinctive abilities, and he probably communicated his experiences with plants to his children and to other members of society. He may have remarked to his friends about, say, the tastiness of a particular plant, or how that plant made him feel after he ate it, or perhaps that he had seen a comrade drop over dead after consuming a particular plant, as surely, many did. And logically, as man developed written communi-

cation, the next step, naturally, was to catalog his experiences in an
orderly fashion for the benefit of those who were to come after. One
of the earliest and best-known of these works was the herbal compiled
by the Greek physician, Dioscorides. It was entitled *De Materia Medica,* and although it was not translated into English until the 17th century, by the famed botanist of Petersfield, England, John Goodyer, it
was the chief source of information for herbalists throughout the Middle
Ages.

Dioscorides' work was illustrated by a Byzantine artist in A.D.
512, who apparently based his work on sketches by Crateus, the personal physician to Mithridates VI Eupator (120–63 B.C.), ruler of Pon-
tus. Mithridates was well versed in the specifics of numerous poisons
and antidotes and was commemorated by having a potent antidote,
mithridate, which was used extensively throughout the Middle Ages,
named for him. Indeed, Dioscorides work was not the first or only
attempt to catalog plants, and, in fact, he complained of the inaccu-
racies of other "modern" writers. He speculated that many writers
were not totally familiar with the appearance of a particular herb
throughout its entire growth cycle, and he attributed their "blunders"
in identification to this unfamiliarity.

But the knowledge of herbs continued to be passed along, in many
ways, first by the Romans, and then throughout Europe as the Roman
armies swept over the land. And it wasn't until the 16th and 17th
centuries that two of the more well-known herbals of England were
produced by John Gerard and Nicholas Culpeper, respectively. About
the same time Gerard's work was being released in England, a Spanish
physician, Nicholas Monardes, was publishing his *Joyful News Out of
the Newe Founde World,* a treatise on herbs which included descrip-
tions and tales of herbs and plants that visitors to the New World had
learned about through contact with the Indians. This was probably
what could be considered the first "American" version of a compendium of herbs, despite the fact that the work was not translated into
English for many years thereafter.

Perhaps the most extensive cataloging and use of herbs in America
was carried on by the Shakers, who were devoted to the natural life
and the natural healing properties of medicinal herbs. They had come
to America from the poorest towns of England to start a life devoted
to peace and freedom from abuse. The Shakers' religious beliefs required them to keep careful records of all aspects of their lives. This
included their extensive use of herbs and plants for medicinal purposes.
Gradually, they started to sell the surplus herbs they had grown to
supply their families, and the scale of their trade in herbs and prepa-

rations eventually grew to be a nationwide enterprise. They documented their research of herbs in the New World, and they learned of new uses and treatments from native American Indians. The Shakers were one of the first American groups to publish a complete catalog of herbs, and they were the first to produce herbs for hospitals and the pharmaceutical market. Their business ventures carried well into the 20th century, and at least two Shaker communities are still in existence at this writing.

Other early Americans, such as Meriwether Lewis and John Clark, documented their experiences—and amazement—with the actions and versatility of native herbs and plants, as taught to them by the Indians while on their historic exploratory trek across the uncharted lands of the western United States. And in the east, the Pennsylvania Dutch also made frequent use of the natural healing properties of medicinal herbs. To this day, they, and especially the Amish, take pride in their self-reliance and utilitarian home remedies made from herbs and medicinal plants, which play a big part in their lives.

Indeed, herbs played a much larger role in the lives of most early families, who needed more help to stay in good physical condition than modern man, because foods at home and those purchased were more likely to be spoiled or tainted than those available today. Without the benefit of today's modern processing, there were more cases of upset stomach and intestinal disorders from eating spoiled food. In addition, less than a century ago there were far fewer doctors and therefore a far greater need for home remedies of all kinds.

But as industry progressed, people had less time to grow their own herbs. More time was devoted to labor, and rapid advances in science and technology made possible the synthetic production of the essence of herbs in great volumes. In doing so, science outpaced the patient garden herbalist. Slowly, people forgot how to use herbs, they had no time to do so, and the sprawling, smoky factories and industrial towns ate up the land, reducing the numbers of those growing herbs at home as they had in the past.

Now, after decades of bombardment with chemical additives and artificial ingredients, people are crying out once again for the "real thing." Growing numbers of people are no longer content to accept foods without knowing the chemicals that are contained in them. But even more than a need to know, many feel that the ability to be self-sufficient at least to some degree has a high priority. In today's depersonalized society, the knowledge and cultivation of herbs can give the individual a feeling of personal achievement and worth.

Chapter 2

Foraging for Herbs

The information in this chapter on foraging was provided by Thomas G. Arehart.

Those who decide to write books specifically on edible or medicinal plants usually go back to the library to gather all of the information possible. With pen (or typewriter) in hand, they proceed to become experts on foraging. Unfortunately for those who are truly interested in knowing more about foraging, most books available are merely repetitions of supposed facts, that is if they are facts at all. And that new book on foraging becomes just another weak collection of age-old speculations or misinterpretations. Developing skills for foraging becomes even more difficult when you discover that some reference texts contain major flaws in even the most recent editions.

How could this have happened? Consider this: One day, in 1884, John Smith bent down and took a nibble of some Belladonna berries. He had heard from a friend that one could eat the ripe, black berries with no harm to body or soul, and, indeed, he subsequently suffered no ill effects. John makes a note of his experience for his forthcoming book on plants of the area. His records, in book form, are published and placed in the library. John, of course, had no idea that this one cursory notation would be perpetuated for many lifetimes.

Years later—say, 1894—a person in search of more knowledge of edible plants for, you guessed it, a forthcoming book, goes to the local library. He finds a book on the subject by one John Smith, and he is amazed to learn that Belladonna berries are edible. Unfortunately, John Smith has since met with an unfortunate demise, and entries in the book have become impossible to confirm. No permissible quantities, no warnings, and no specific time of the year when the tasty berries might be edible were given in the text. Thus, compiling data instead of experience, this new fellow duplicates the original material, with minor changes to avoid plagiarism. This 1894 edition simply states: "Belladonna—berries edible when ripe."

Many years pass, and a graduate student in 1934, who wants a complete listing of edible plants for a thesis, goes to her college library to check for references. She finds two publications saying that Belladonna berries are edible, and, in a rush to complete her paper on time, she records: "Belladonna—berries eaten since 1884."

Then, in 1954, a person who has heard of the wonders of the outdoors, tries to discover what he, too, can write on the subject of edible plants. He reads that Belladonna berries, when ripe, are edible. Well, everyone knows that *any* berry is only edible when it is ripe!! So, in the 1954 version of his new book appears this surprisingly simple account: "Belladonna berries are edible."

So, in 1984, a young lad who has only recently become aware of the joys of wild plants and foraging in the wilderness, takes out his trusty edible plant handbook as he comes upon a new plant. To his delight, he reads that the juicy black berries he sees before him are "edible." He finds them quite tasty, but in only a few moments he is rendered totally helpless, lying on the ground in a state of violent illness.

This rather extended story is meant as a vehicle with which to demonstrate how misleading some information found in herbal and foraging guides can be, and how it got that way. Remember that the original information comes usually from just a single experience and may become so distorted or abridged that it becomes almost a fatal mistake to believe it. Often, no permissible amounts to be consumed are given, the description of the plant is insufficient for proper identification, and the necessary antidote in cases of accidental poisoning are left out. All of the above are necessary before attempting to consume a heretofore toxic plant.

Information provided in this text includes, for appropriate species, the areas in which they grow, their habitat, and how to collect them. Listings on so-called edible plants come, not from a book, but from the pages of experience—and then, not only from my experience, but that also of the countless number of students I have taught on the trail, in the back of college campuses, and in rural yards. My survival, and that of the several students who regularly accompany me, depends upon my knowledge and proper use of several important guidelines. And the compilation of this information and these tips has been a ten-to-fifteen-year effort and the result of as many years of teaching and experiencing foraging.

Why Are You Foraging?

Have you ever read of the glories of eating Skunk Cabbage? It is a marvelous little plant that, after hours and hours of boiling and cooking, will still taste and smell like burning rubber—proving that there are many tastes out in the wild that should be completely avoided. Even on the roughest survival trips, I never suggest this plant except as a joke or to gain revenge. So, as you can see, the term "edible" is so broad and general that even a disgusting plant like Skunk Cabbage can be listed as "edible," as can the paper this book is printed on. All edible plants are edible, but that does not guarantee that you can use them for food or for medicine. What you are looking for is nutritional value and specific chemical constituents.

Ten Golden Rules

There are guidelines that need to be set so as to prevent a beginning forager from destroying both nature and his life. You don't just pick up a list of edible plants and go out collecting them: there are a few things to know. The following rules are not engraved in stone, so to speak, but they are a look into the destructive powers found within you, as well as the destructive powers that await you. For every rule you read here, I relate a story of how I broke it and what was lost. The punishment for breaking these rules was that a small part of nature was lost, or that my life was placed in jeopardy. These rules will hopefully bring safety and peace into your foraging. Regard them as a look into the costly mistakes that I have made, and understand that you need not repeat them for yourself.

Rule 1—*For every 12 plants you find, pick only one.*

Once, when I was also beginning, I found, with the guidance of an expert botanist, a small patch of Indian Cucumbers. I was told that the roots were not only edible, but were much sought after by those in the know. Putting the advice to the test, I picked up a small root and proceeded to taste it. I agreed: the juices were sensational. Returning home, I told my friends, who knew very little about the subject, of the new taste I had just experienced. They showed much enthusiasm to try it for themselves, and soon we all were standing beside the small patch, digging away. One by one, everyone sampled the roots and went digging for more. In a matter of minutes, we had eaten the entire patch. Not one plant was left to grow, to reproduce, or to return the next year. We not only destroyed a part of nature, but also the composure of my botanist friend.

The only plant, as recorded in scientific journals, that keeps increasing in number year after year is the infamous Dandelion. All other plants are declining in number. The botanist I had betrayed told me a biological rule: if you come upon a small patch of plants, as I had, you should only take or destroy one plant for every twelve that are there. This insures the survival of the plant so that you may return to that patch year after year. Going back to the area where those Indian Cucumbers once existed would now be a waste of time for me. Thus, breaking this rule becomes a punishment unto itself for those who care.

Rule 2—*Avoid eating roots.*

The previous rule may well serve as the best example for this one; but perhaps an additional story will provide another dimension.

Taking weekend walks with friends and other interested students affords me the opportunity to explore those areas I would otherwise never get a chance to see. On one such walk, we, as a group, came across a rather large patch of Fragrant Water Lilies. We went ahead, collecting the roots to be transplanted in my guests' pond. The roots were edible, in addition, so we decided to collect several more to eat for lunch.

Later that day, from some local residents, we learned that the pond we collected from earlier that day was quite unique. It *had* contained the only naturally growing patch of these Water Lilies in not only the state, but on the entire East Coast. We could return those plants meant for transplanting, but not those already spread out on our lunch table.

When you eat the roots of a plant, it is immediately killed. For this reason, the listings in this book explain the multiple uses of leaves, stems, berries, and seeds that can provide you with years of delight. When you read that only the roots of a plant are useful, don't exploit the plant unless it happens to be the Dandelion. Your survival is one thing, but common sense is just as important. You can't realize how rare and valuable some of these plants are until you go back to your favorite patch only to find that it has been destroyed by novice foragers.

Rule 3—*Know mimicry plants.*

For every edible plant, nature has deviously provided one, two, or sometimes three look-alike plants. Some look-alikes are harmless; others are deadly. Mistake Yarrow for Wild Carrot, and you'll receive an aromatic gift for your error. Mistake Water Hemlock for Wild Carrot and you may have a long stay in the hospital. If you are not so fortunate to go on an outing with a friend who knows mimicry plants, take time to get to know those plants you are looking for thoroughly. Know their parts, where they are found, and the major differences between them and their cousins. I cannot stress enough the need for correct identification of the plants you are using. The information found in any text on foraging is useless until you can identify the plants you're looking for correctly and safely.

Rule 4—*Know when the plants are edible.*

The most discouraging hunts I have ever been on were those on which I went looking for a particular plant on the recommendations of some book and discovered with much dismay that the plant had been dead for about two months. Mayapples, for example, cannot be collected in the fall. And there's no use looking for Rose hips in early spring. A list of delicious plants is useless if they're dead or just sprouting when you find them. Know when your plants are edible!

Rule 5—*Know your interference conditions.*

One summer day, I was showing a group of students the wonders of the outdoors. We came upon a small cluster of Cattails. No one in the group knew how or when to eat these little vegetables. Bending down, I grabbed a rootstock, cleansed it, and bit a small chunk off. Normally the rootstock is starchy and somewhat sweet, but this time it was bitter and toxic. In less than five minutes, I was on the ground, dizzy, steadily growing weaker, and with a burning sensation in my mouth!

The problem here wasn't obvious until I took some time to recover and look around. While it is true that Cattails are normally edible, the one I bit into was definitely poisonous. Why? Upstream, I found the answer. You see, Cattails are aquatic plants, but so are the toxic Marsh Marigolds, which contain a water soluble toxin, palustrin, that can be quite dangerous if taken in large doses. These plants were growing just a few yards upstream from the Cattails. Laboratory analysis showed that these Cattails picked up the poison from the Marigolds and concentrated it in their roots. Eating them for an extended period of time might have proved deadly.

Look around before you eat your plant. Look for generally dangerous conditions that are found normally. Is the stream stagnant? Are there any poisonous plants around? How far away, or how close, are they? Are there any dead or dying plants or animals nearby? Be observant. You can't expect to survive long if you concentrate on looking for one plant without respect to those growing nearby.

Rule 6—*Know your absorbance conditions.*

While interference is due to natural causes, absorbance is due to completely unnatural causes. And while one plant may "interfere" with the edibility of another, foreign chemicals will not only affect edibility but may also create toxicity. With camping and hiking on the upsurge, more people insist on dragging along cans, aluminum foil, and other trash, and these items are found in even the most barren areas these days. Once, while camping in one of these areas, I found a most disturbing sign. Local Boy Scouts had posted a notice that the drinking water in the area was contaminated. The notice explained that some previous campers had buried their garbage above the water source unwittingly, and as it decomposed it found its way into the drinking water. No plants remained, not even the simplest algae, in the stream. The water had to be treated chemically or boiled before use. And the surviving plants growing around the stream, which received varying smaller doses of the poison, had, at any rate, absorbed some of the decomposed material. Any person who ate portions of these plants would probably have received small amounts of toxic metals that, as medical texts show, accumulate irreversibly in the human body. And those plants would probably not have shown any signs of their dangerous contamination.

Rule 7—*Know the yellow rule.*

Travelling many miles in search of one plant may seem a bit eccentric to many people. And, while you normally don't have to walk several miles to find edible plants, you may eventually get picky after you know what is out there to eat and enjoy. But there is yet another reason for turning down your favorite plant, even after you have trekked forever to find it. You may find that your favorite plant is broken and yellowing and perhaps even dead. Much like the vegetables that sit on your supermarket's shelves for extended periods, these yellowed plants may lose much of their nutritive value. So, even though you've come a long way, it's best to avoid a sickly plant if at all possible. Walk on a little further, if need be.

Rule 8—*There is no such thing as "just" plants.*

I felt very confident and thought myself smart by my second year of taking people out on foraging trips. Once, a student in one of these "classes" asked what a certain plant was. I looked for a second, and said, "Oh, that's just Wild Licorice. It grows everywhere. Let me show you how delicious the seeds are!" I then proceeded to go over to the plant and show the student how easy it is to poison yourself with the seeds of Poison Hemlock. There is no such thing as just Wild Licorice, or just Wild Ginger, or even just Wild Dandelion. Though, at this point in time, my body is now adapted to many forms of toxic wild elements, it was still a serious mistake to assume that any plant could be labeled "just."

Rule 9—*Don't try to transplant.*

Taking frequent trips to the wild mountains of Tennessee allowed me to see the progressive expansion of areas in which a certain plant was growing. It was like seeing an old friend growing bigger and better. But one year, I returned and found that a cluster of Pink Lady's Slippers had been moved. I asked a state park guide why, and I was informed that many tourists had been pilfering the pretty plants for transplanting in their home gardens.

What these novice gardeners failed to realize was that, for such a marvelous bloom to appear on Pink Lady's Slipper, it requires seven years of growth, along with at least nine different types of soil fungi to provide the proper nutrients. So the plants they would transplant would only die; and instead of leaving these old friends in a place where they could be seen and enjoyed every year, these people, in one selfish, albeit unintentional act, killed them. So please, no matter how thrilled you are by the taste, smell, or use of these wild plants, leave them where they will probably grow best: in their native soil.

Rule 10—*Every plant has a use.*

Some plants are edible; some plants are medicinal. And some plants have utility, such as for the making of bedding or clothing. Some

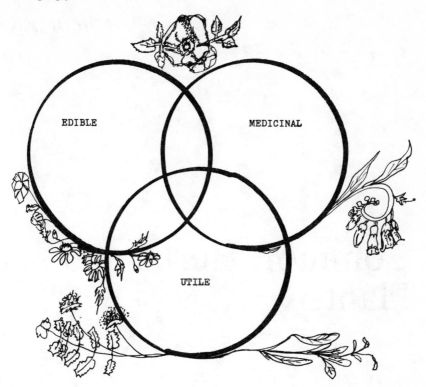

plants are both edible and medicinal; some plants are both medicinal and utile. Few plants are edible, medicinal, and utile; but *none* are useless. Take a glance at the chart I have drawn, which shows the three functions of plants. One circle represents the edible plants, one circle represents the medicinal plants, and another circle represents the plants with some utility. As you can see, few plants fit all three categories, but all plants have some use. The challenge of the modern forager is to find that use.

Not all of the following 200 or so herbs and plants can be found growing wild in North America. Some you'll have to grow for yourself. But whether you grow them or forage for them, armed with the previous advice, your experiences in the wild, at least, should prove to be both pleasurable and fascinating.

Chapter 3

Common Herbs and Plants

To help you in finding a particular herb in this book, we have provided the following listing of the herbs as they are arranged in this chapter.

ARISTOLOCHIACEAE, The Birthwort Family
Asarum canadense .... Wild Ginger

ASTERACEAE, or COMPOSITAE, The Sunflower or Daisy Family
Achillea millefolium .... Yarrow
Antennaria neodioica .... Pussytoes
Anthemis cotula .... Mayweed
Anthemis nobilis .... Chamomile
Arctium minus .... Common Burdock
Arctium lapa .... Great Burdock
Artemisia abrotanum .... Southernwood
Artemisia absinthium .... Wormwood
Artemisia drancunculus .... Tarragon
Artemisia tridentata .... Big Sagebrush
Artemisia vulgaris .... Mugwort
Calendula officinalis .... Marigold
Carthamus tinctorius .... Safflower
Centaurea cyanus .... Cornflower
Centaurea maculosa .... Spotted Knapweed
Chrysanthemum leucanthemum .... Oxeye Daisy
Cichorium intybus .... Chicory
Cnicus benedictus .... Blessed Thistle
Helianthus tuberosus .... Jerusalem Artichoke
Inula helenium .... Elecampane
Solidago odora .... Sweet Goldenrod
Solidago tenuifolia .... Slender Fragrant Goldenrod
Tancetum vulgare .... Common Tansy
Taraxacum officinale .... Common Dandelion
Tussilago farfara .... Coltsfoot

BALSAMINACEAE, The Touch-Me-Not Family
Impatiens capensis .... Jewelweed

BERBERIDACEAE, The Barberry Family
Podophyllum peltatum .... Mayapple

BORAGINACEAE, The Borage Family
Anchusa officinalis .... Alkanet
Borago officinalis .... Borage
Myosotis scorpioides .... Forget-Me-Not
Symphytum officinalis .... Comfrey

BRASSICACEAE, or CRUCIFERAE, The Mustard Family

Barbarea vulgaris	Common Winter Cress
Brassica alba	White Mustard
Brassica nigra	Black Mustard
Capsella bursa-pastoris	Shepherd's Purse
Cardamine bulbosa	Spring Cress
Cardamine pensylvanica	Pennsylvania Bittercress
Isatis tinctoria	Woad
Lepidium sativum	Garden Cress
Nasturtium officinale	True Watercress
Sinapis arvensis	Field Mustard
Sisymbrium alliaria	Garlic Mustard
Sisymbrium officinale	Hedge Mustard

CACTACEAE, The Cactus Family

Opuntia humifusa	Prickly Pear
Opuntia fiscus-indica	Quesode

CAPRIFOLIACEAE, The Honeysuckle Family

Sambucus canadensis	Elderberry
Sambucus nigra	Black Elderberry

CARYOPHYLLACEAE, The Pink or Carnation Family

Cerastium vulgatum	Mouse-Eared Chickweed
Gypsophila struthium	Egyptian Soapwort
Saponaria officinalis	Bouncing Bet
Stellaria media	Starwort Chickweed
Stellaria pubera	Star Chickweed

CHENOPODIACEAE, The Goosefoot Family

Chenopodium album	Lamb's Quarters
Chenopodium ambrosioides	Mexican Tea
Chenopodium bonus henricus	Good King Henry
Salicornia europaea	Slender Glasswort
Salicornia herbacea	Jointed Glasswort

CRASSULACEAE, The Sedum, or Stonecrop Family

Sempervivum tectorum	Houseleek

DIPSACACEAE, The Teasel Family

Dipsacus sylvestris	Teasel

ERICACEAE, The Heath Family

Gaultheria procumbens	Wintergreen

GERANIACEAE, The Geranium Family
Geranium maculatum	Wild Geranium
Geranium robertianum	Herb Robert

IRIDACEAE, The Iris Family
Crocus sativus	Saffron
Iris versicolor	Iris
Iris germanica florentina	Orris

LABIATAE, or LAMIACEAE, The Mint Family
Betonica officinalis	Wood Betony
Cunila origanoides	Dittany
Hedeoma puleogioides	American Pennyroyal
Hyssopus officinalis	Hyssop
Lavandula officinalis	English Lavender
Lavandula spica	Spike Lavender
Lavandula stoechas	French Lavender
Mentha piperita	Peppermint
Mentha pulegium	English Pennyroyal
Mentha spicata	Spearmint
Monarda didyma	Bee Balm
Monarda fistulosa	Wild Bergamot
Nepeta cataria	Catnip
Ocimum basilicum	Basil
Origanum marjorana	Sweet Marjoram
Origanum vulgare	Wild Marjoram
Origanum mexicana	Oregano
Prunella vulgaris	Woundwort
Rosmarinus officinalis	Rosemary
Salvia azurea	Blue Salvia
Salvia coccinea	Salvia
Salvia mellifer	Black Sage
Salvia lyrata	Wild Sage
Salvia officinalis	Sage
Salvia sclarea	Garden Clary
Satureia hortensis	Summer Savory
Satureia montana	Winter Savory
Thymus serpyllum	Wild Thyme
Thymus vulgaris	Garden Thyme
Teucrium canadense	Germander

LAURACEAE, The Laurel Family
Laurus nobilis	Bay

LEGUMINOSAE, or FABACEAE, The Bean or Pea Family

Cassia marilandica	Wild Senna
Glycyrrhiza glabra	Licorice
Melilotus alba	Melilot
Melilotus officinale	Yellow Sweet Clover
Medicago sativa	Alfalfa
Trifolium pratense	Red Clover
Trifolium repens	White Clover
Trigonella foenum-	Fenugreek
graecum	

LILIACEAE, The Lily Family

Aletris farinosa	Stargrass
Allium canadense	Wild Garlic
Allium cepa	Onion
Allium cepa, var.	Egyptian Onion
aggregatum	
Allium sativum	Garlic
Allium schoenoprasum	Chives
Allium tricoccum	Wild Leek
Asparagus officinalis	Asparagus
Convallaria majalis	Lily-of-the-Valley
Erythronium albidum	White Dog-Tooth Violet
Erythronium americanum	Adder's Tongue
Erythronium propullans	Minnesota Adder's Tongue
Medeola virginiana	Indian Cucumber Root
Polygonatum biflorum	Solomon's Seal
Smilax ornata	Sarsaparilla
Veratrum viride	Hellebore
Yucca aloifolia	Spanish Bayonet
Yucca glauca	Soapweed
Yucca filamentosa	Yucca

MALVACEAE, The Mallow Family

Abutilon theophrasti	Velvetleaf
Althaea officinalis	Marshmallow
Hibiscus coccineus	Hibiscus
Malva neglecta	Cheese
Malva rotundifolia	Low Mallow
Malva sylvestris	High Mallow

ORCHIDACEAE, The Orchid Family

Cypripedium acaule	Pink Lady's Slipper

PLANTAGINACEAE, The Plantain Family
Plantago major	Common Plantain
Plantago lanceolata	English Plantain

POLYGONACEAE, The Buckwheat Family
Polygonum aviculare	Prostrate Knotweed
Polygonum bistorta	Bistort
Polygonum pensylvanicum	Pennsylvania Smartweed
Polygonum persicaria	Lady's Thumb
Rumex acetosa	Garden Sorrel
Rumex acetosella	Sheep Sorrel
Rumex alpinus	Monk's Rhubarb
Rumex crispus	Curly Dock
Rumex hydrolapathum	Great Water Dock
Rumex obtusifolius	Bitter Dock
Rumex scutatus	French Sorrel

PORTULACACEAE, The Purslane Family
Portulaca oleracea	Purslane

RANUNCULACEAE, The Buttercup Family
Aconitum uncinatum	Wolfsbane
Aconitum napellus	European Wolfsbane
Aquilegia canadensis	Columbine
Hydrastis canadensis	Golden Seal

ROSACEAE, The Rose Family
Agrimonia gryposepala	Agrimony
Agrimonia eupatoria	European Agrimony
Fagaria virginiana	Common Strawberry
Fragaria vesca	Wood Strawberry
Potentilla simplex	Common Cinquefoil
Potentilla canadensis	Canadian Dwarf Cinquefoil
Potentilla reptans	Five-Finger Grass
Potentilla anserina	Silverweed
Rosa damascena	Damask Rose
Rosa multiflora	Multiflora Rose
Rosa virginiana	Virginia Rose
Rosa rugosa	Rugosa Rose
Rosa eleganteria	Sweet-Briar Rose
Rosa canina	Wild Dog Rose
Rubus strigosus	Red Raspberry
Sanguisorba poterium	Salad Burnet
Sanguisorba canadensis	Canadian Burnet

RUBIACEAE, The Bedstraw Family
    Asperula odorata   Sweet Woodruff
    Galium verum   Yellow Bedstraw

RUTACEAE, The Rue Family
    Dictamnus albus   Gas Plant
    Ruta graveolens   Rue

SCROPHULARIACEAE, The Snapdragon or Figwort Family
    Digitalis purpurea   Foxglove
    Euphrasia americana   Eyebright
    Linaria vulgaris   Yellow Toadflax
    Verbascum thapsus   Great Mullein
    Veronica officinalis   Speedwell

SOLANACEAE, The Nightshade Family
    Atropa belladonna   Belladonna
    Physalis alkekengi   Chinese Lantern Plant
    Physalis heterophylla   Clammy Ground Cherry
    Solanum americanum   Common Nightshade
    Solanum dulcamara   Climbing Nightshade

TROPAEOLACEAE, The Nasturtium Family
    Tropaeolum majus   Nasturtium

UMBELLIFERAE, or APIACEAE, The Carrot or Parsley Family
    Anethum graveolens   Dill
    Angelica archangelica   Angelica
    Anthriscus cerefolium   Chervil
    Carum carvi   Caraway
    Coriandrum sativum   Coriander
    Cuminum cyminum   Cumin
    Daucus carota   Wild Carrot
    Foeniculum vulgare   Fennel
    Levisticum officinale   Lovage
    Osmorhiza claytoni   Sweet Cicely
    Pimpinella anisum   Anise
    Petroselinum crispum   Parsley

AMARANTHACEAE
The Amaranth Family

Members of the Amaranth family are mostly herbs whose flowers bloom in spikelike clusters; they are usually somewhat inconspicuous, however. The flowers are usually small and crowded together along with prickly bracts. The flowers are unisexual or bisexual, with five dry and thin sepals. There are no petals. There are usually five or less stamens. The calyx is often a bright color and sometimes scaly. All flower parts are attached at the base of the ovary. The fruit is one-seeded. Amaranths are generally found in warm regions of the North American continent. The leaves of members of Amaranthaceae are usually simple, alternate, or opposite.

PRINCE'S FEATHER
Amaranthaceae
Amaranthus hypochondriacus

The densely packed crimson flower spikes of this Amaranth secure its place in many a garden, but it is also found in the wild in the central

PRINCE'S FEATHER
Amaranthus hypochondriacus

part of the United States. Prince's Feather blooms in August. The plant grows to four feet in height and has a sturdy, branching stem. The leaves are oblong with a reddish-purple splotch.

Prince's Feather is astringent and, in the past, has been used to assuage diarrhea, hemmorhage from the bowel, and excessive menstruation. A wash made from the herb will supposedly help skin irritations, and a gargle is said to soothe mouth and throat irritations.

GREEN AMARANTH, Pigweed
Amaranthaceae
Amaranthus retroflexus

Hardly an attractive herb, Green Amaranth is quite commonly expelled from yards as a common weed. Often growing to more than four feet in height, the rough, hairy stems form a spikelike cluster of green flowers. Mixed in between the flowers are bristlelike bracts. Flowers appear from August to October. The long, oval leaves grow from three to six inches in length and are commonly stalked. Green Amaranth will grow just about anywhere in waste areas or in your garden.

Birds are attracted to the numerous seeds of this herb. Indians ate

GREEN AMARANTH, Pigweed
Amaranthus retroflexus

Green Amaranth as a green in salads or in prepared form like spinach. It was one of more than fifty herbs used in a similar fashion.

THORNY PIGWEED
Amaranthaceae
Amaranthus spinosus

Used as a diuretic in India, Thorny Pigweed is commonly found in America and is known for its attractiveness to songbirds. They like the large volume of seeds that forms once the tiny green flowers fade. The male flowers grow in terminal spikes up to six inches long, while the female flowers form in round clusters that are intertwined with bristlelike bracts as long as the sepals. The flowers appear atop one of the many branches. The stems are reddish in color, especially near the ground where they form the roots. Thorny Pigweed grows to four feet in height. The leaves, like many of the Amaranth family, are mucilaginous and contain some degree of sugar.

The leaves are said to be useful as potherbs and can be cooked like Spinach or chopped and added to salads. Thorny Pigweed will grow anywhere outdoors, but its appearance negates its practicality as an indoor plant.

THORNY PIGWEED
Amaranthus spinosus

APOCYNACEAE
The Dogbane Family

The herbs or shrubs of this colorfully named family produce a milky juice. Flowers appear in a cluster or by themselves. The flowers are symmetrical: there are five petals and the same number of stamens and sepals. All are attached at the base of the ovary. The leaves are opposite, whorled, or alternate, and simple in form. There are said to be about 200 genera and ten times as many species. The seeds are often hairy and are comprised of two follicles.

PERIWINKLE
Apocynaceae
Vinca minor

This garden escape bears a sweet scent in the spring when its sky-blue flowers are in bloom. The flowers appear from April to May. The five-lobed flower grows alone in the axil of the leaf, but this just tends to show off the lovely wide blue petals, which, upon close inspection, show a white star in the center. Periwinkle grows on a trailing vine

PERIWINKLE
Vinca minor

that remains green all year round, and grows to a height of eight inches. The leaves are about an inch and a half long and grow opposite each other on the stem. When in good health, the dark green leaves have a shiny appearance.

Periwinkle is a great ground cover under trees. The genus name, *Vinca,* comes from the Latin *vincio,* meaning "to bind," a reference to the herb's entangling characteristics. Periwinkle was a favorite herb for making love potions, and it was said to dispel evil spirits. It was used to make garlands for the Italians, who used the garlands to adorn the bodies of young children who had died. Other Europeans thought it was a flower of immortality or friendship.

Periwinkle is said to be an excellent astringent and tonic. It is also reputed to be a good remedy for cramps and nosebleeds. Periwinkle jelly is thought by some to stave off nightmares. An ointment of the bruised leaves combined with lard is good for skin irritations and piles, supposedly.

Under trees, along banks, and along roadsides throughout North America, Periwinkle makes a most pleasant addition to the shaded areas of your garden. The drifting scent from the spring flowers is sure to perk up any winterladen spirit. Propagate by root divisions or cuttings early in the spring or late in the fall. Water the plants well after the cuttings have been set.

ARACEAE
The Arum Family

Frequently, members of the Arum family are herbs and climbers to boot. The plants grow to various sizes and in different habitats. The leaves are quite large, and are either simple or compound, with net-veining. These herbs have a petiole that acts as a sheath. A spike of flowers grows on a fleshy stalk or a spathe, which is often a large, showy bract that folds around the flower spike. The flowers are unisexual and have parts attached at the base of the ovary, but they are free from it. There is one pistil that is made up of one or more carpels that usually forms a berry. Most of the species of this family are tropical, and there are supposedly about 100 genera and around 2,000 species.

JACK-IN-THE-PULPIT
Araceae
Arisaema triphyllum

Recognizing this herb in the wild is easy. Jack-in-the-Pulpit has a distinctive flower which has a part resembling a hood, which surrounds

JACK-IN-THE-PULPIT
Arisaema triphyllum

a greenish-yellow, fleshy erect spike. The hood is a green or purplish-brown color, and may possess streaks or splotches on its outer surface. The main stem may contain two three-part leaves that are quite large and deeply veined. This perennial has been known to attain a height of three feet, and it produces a tight cluster of bright red berries in the late summer. The root, or corm, of the plant is its most practical part; Jack-in-the-Pulpit is found in wet areas from southern Canada through the eastern United States. It is also said to be present in areas of South America.

The root of Jack-in-the-Pulpit, when raw, has an extremely biting taste, and will cause severe irritation of the mouth and tongue if chewed. Cooking the root, or corm, is said to eliminate that unpleasant trait, and it is said that the American Indians gathered the roots of Jack-in-the-Pulpit regularly for use as a vegetable. Other Indian tribes are said to have used the roots in various forms to relieve headache and to serve as a form of contraception.

SKUNK CABBAGE
Araceae
Symplocarpus foetidus

Unlike many plants that blossom in the spring with a sweet-smelling flower, Skunk Cabbage, as its common name implies, has a hideous

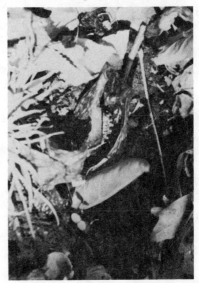

SKUNK CABBAGE
Symplocarpus foetidus

smell. Pushing the snow away in the earliest of spring, this herb's brownish-purple and green-mottled spathe plows its way to the surface, melting the snow with the heat generated by its own growth. A large tuft of green leaves shoots up beside that blossom, unfurling eventually to cover a large area. Those leaves can each be a foot wide and up to two feet long; they have heavy stalks and are deeply veined. The flower is comprised of one leaf or spathe, which can reach a height of six inches. The herb is found in southern Canada and the northern United States in swamps and wet wooded areas.

The odor we find so displeasing is very attractive to insects, who create quite a fuss pollinating the Skunk Cabbage.

Herbalists say the leaves can be eaten as greens, but the resulting smell of cooking hardly makes the effort worthwhile. There are some alleged medicinal virtues, however.

The root is powdered and sometimes added to honey to form a medicine that will calm the bronchial system and help the asthma or whooping cough victim. Do not eat the fresh plant, however: it has acrid properties. If ingested in sufficient quantity, it can cause vomiting and headache.

ARALIACEAE
The Ginseng Family

Members of the Ginseng family can be herbs, shrubs or sometimes trees that bear alternate or compound or divided leaves at the base and are frequently sheathed petioles. The flowers of species of this family are symmetrical, and they form heads or umbels. They are unisexual or bisexual, and have five stamens, one pistil, two to five carpels, and a matching number of styles. The fruit is a berry or a drupe. There are a speculated 65 genera and perhaps as many as 750 species. Many grow in northern climates, but most of them are found in the tropics.

WILD SARSAPARILLA
Araliaceae
Aralia nudicaulis

A greenish-white cluster of flowers nods beneath large, umbrellalike leaves in this species. The leaves come out in three parts, and they have toothed edges. There are three to five leaflets in each leaf.

WILD SARSAPARILLA
Aralia nudicaulis

The whole plant grows to a height of fifteen inches. It flowers in July and August in upland, woody places from British Columbia, south to Georgia, and west.

The species name comes from two Latin words, *nudus,* and *cauli,* which, when translated, mean "naked stalk." It is a reference to the flower stalk, which bears no leaves. There are jagged pricklies all over this plant, especially on the thick, square stems, which one must be careful of when trying to harvest the root. The root is in the form of a rhizome, and it can be crushed and made into a tea. But the herb is probably more well known in the form of the popular soft drink, Sarsaparilla, which rocketed to fame in the early 1960s because of the much-watched TV Western, "Sugarfoot." Sugarfoot, the hero and namesake of the show, would, once a week, walk into the rough-and-tumble saloon of some unidentified Western town and order a frothy mug of Sarsaparilla. The plant was said to have great medicinal value among the Indians.

Wild Sarsaparilla tea was said to be an excellent spring tonic and blood purifier. It would cure upset stomachs, the common cold, and aching joints, according to Indian researchers. The roots, when pulverized, were used on skin blemishes and wounds in the form of a poultice. The Indians held it to be effective in the fight against venereal disease. The oil of the root of the herb was dropped into the ear to cure earache and deafness.

Wild Sarsaparilla prefers partial shade and rich, loamy soil that is well-drained. Start your plants from seeds or root cuttings in the usual manner.

GINSENG
Araliaceae
Panax quinquefolium

The mystical Ginseng is a handsome plant of moderate size, found in the shady woodlands of North America and other parts of the world. You'll recognize it by its three large compound leaves that rise from the moist floor of the woods. The leaves form a circle around the stem. Each leaf cluster bears five leaflets and can be up to a foot long. The leaves are toothed and form a point at the end. The plant has been known to reach three feet in height, but for the most part it does not exceed two feet. The quality of soil makes the difference. The flowers rise from the center of the leafstalks from May to July. They're greenish

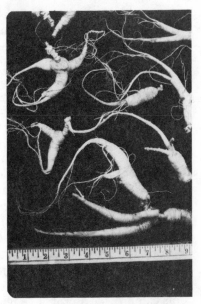

GINSENG
Panax quinquefolium

GINSENG ROOTS
Panax quinquefolium

and white, and sometimes have a yellow cast, with a sweet, faint scent like Lily-of-the-Valley. They're quite tiny and have five petals. When they fade away, a cluster of ripe red berries forms, each berry containing two seeds.

Ginseng has long been regarded as the ultimate cure-all of the Orient. Among its powers is its supposed ability to excite sexual desire. The common name is derived from the Chinese *Jin-chen,* which means manlike. The name is a direct reference to the fleshy root that is often found growing in two or more forks. The more the root resembles a man in appearance, the more valuable it is considered to be. The Orientals think it then has stronger medicinal properties. Roots of Ginseng found in the wild have a much higher value commercially, because the always discriminating buyers of this herb will refuse any cultivated variety. The root covering of Ginseng plants grown in the wild has an elephant-skin-like texture, with visible concentric rings, while the roots of cultivated varieties do not.

From hundreds of years of use, the Chinese have learned from their forefathers the valuable tonic properties attributed to this herb. Ginseng is said to make the drinker of its tea live longer, fill the heart with gladness, and cure just about anything that ails you. The famous tea is made by grinding the dried roots into a powder. Three ounces of that powder, combined with one ounce of sugar, and sixty drops of Wintergreen oil makes the basic mixture. A teaspoon of that is placed in a ceramic cup, and boiling water is poured over it. Let that mixture steep for ten minutes, and drink it as hot as you can bear. Do this with each meal, and it is said that you will add at least ten years to your life. Ginseng tea is a great substitute for coffee.

You must grow this herb in a rich soil with plenty of shade. Ginseng does not transplant well, but to do so, select a place under the dark, leafy boughs of trees, and use a rich compost. Dig a large hole and set the root into it with the stem side up. Fill in the dirt around it, and tamp down. Then, snip off the stem just above ground level. If you're lucky, and do not let the root become too dry or too moist, the Ginseng should send up a shoot in the fall or the spring. It is said to grow quite well with Golden Seal. To start Ginseng, press one of the scarlet-red berries from the fruit cluster into the moist soil with your finger. It takes a long year to eighteen months to germinate and requires five years of growth to make a harvestable root. In sufficient quantities the roots can provide a profitable bounty. Some export houses have been known to pay 160 dollars for light, dry Ginseng roots. You can tell the age of a Ginseng plant in two ways: by the number of nodules on the neck of the root, which indicates how many stems; and thus, seasons, have been grown and withered (it is important to harvest plants that are at least five years of age for the root loses 80 percent of its weight in drying); and by the number of leaves and the appearance of the fruit. A first-year plant will hav only one leaflet. By the second year two leaflets emerge, and in the third season three leaves of five leaflets each, forms along with the first seed pod. In the fourth year, four leaves appear. It cannot be stressed too much that Ginseng is a sensitive plant that, since it grows so slowly, must be treated with respect. It is said that, for each root of the Ginseng you gather, seven seeds should be sown in its place.

ARISTOLOCHIACEAE
The Birthwort Family

Members of the Birthwort family are usually herbs, but they can also be woody vines. The leaves of these plants are often heart-shaped: the flowers unusual and rather large. The flowers are usually found with no petals, and the calyx folds over. It is made of three brownish-purple parts. Six or more stamens are found within, and all parts are attached at the top of the ovary. The untoothed leaves appear in an alternate pattern and are either stalked or basal. The fruits of these plants are comprised of a capsule with four to six chambers. This is a small family with six genera and a few hundred species, some of which are cultivated.

WILD GINGER, Canada Snakeroot
Aristolochiaceae
Asarum canadense

Wild Ginger is a pleasant looking little plant with two velvety kidney-shaped leaves. It grows to about a foot in height, and the leaves

WILD GINGER
Asarum canadense

can be from four to eight inches in width. A solitary brownish-purple flower grows at ground level between the two leaf stalks. It is somewhat bell-shaped, drooping, and has three pointed lobes. Wild Ginger flowers from April through May. Its fruit is a leathery six-chambered capsule. The plant is a perennial, and it is commonly found along roadsides and in rich wooded areas. The valuable rootstock of this herb is a yellowish color and jointed.

Wild Ginger root has a ginger odor and taste, and it can, therefore, be used as a substitute to make ginger ale. A strong decoction of the root has been used by some Indian tribes as a contraceptive. And the root, when powdered, is extremely effective in inducing sneezing; so it has been recommended for treatment of head colds. Wild Ginger root in various forms has been used in the past to treat throat afflictions, dropsy, and intestinal disorders. But while it is said to be good for bowel spasms, it is also an extremely powerful purgative when ingested in sufficient quantity. The root is best gathered in early autumn for peak potency.

ASTERACEAE or COMPOSITAE
The Sunflower or Daisy Family

This family contains mostly herbaceous, but often woody, plants with both simple and compound leaves. The leaves are both alternate and opposite. The distinctive trait of the members of this family is the flower head, which looks like a large, symmetrical flower that is cupped in a circle of green bracts. The plants of Asteraceae prefer a cool, temperate climate, and the family comprises one of the largest groups of plants, with genera numbering over 900 and about 20,000 species. The fruits, which often provide a valuable food source, are one-seeded, and have a hard shell.

YARROW
Asteraceae
Achillea millefolium

Yarrow is an exceedingly common perennial in old fields, meadows, and along roadsides. It is recognized by its tiny creamy-white or pinkish flowers, which form small, flat-topped clusters atop the whitish,

YARROW
Achillea millefolium

hairy stem. The leaves of the plant are light, feathery, and somewhat fernlike. They tend to clasp the stem at their bases and grow to a length of four inches. This herb is found everywhere; it grows to a height of three feet, and the flowers appear from June through September.

Yarrow is best known for its medicinal applications, especially as a treatment for wounds and nosebleeds. The Greek warrior Achilles was supposed to have nursed his soldiers' wounds with the herb; hence the genus name. Beyond that, herbalists seem to contradict each other when discussing the characteristics of Yarrow. Some agree that the herb will stop bleeding, while others maintain that the leaves of the plant, when brought in contact with the nose, will provoke an immediate nosebleed. Nonetheless, scientific chemical analysis proves that Yarrow does, indeed, contain chemical constituents, such as *achillein,* and achilleic acid, which do have styptic and somewhat astringent properties. An extract of the leaves and flowers, in equal parts, was recommended as the basis of a stimulating tonic. Yarrow, Borax, and Chamomile water, when mixed, were recommended at one time as an agent to bring on menstruation.

Yarrow is a great plant for beginning gardeners, because it will grow with little attention and varying amounts of sun. The roots may tend to spread out, though; but that is of little consequence, since it is reputed to enhance the flavor and aroma of any other surrounding herbs and will help them ward off disease as well as keep beetles, flies, and ants away. For this reason, Yarrow is often cultivated near the house. Yellow and orange varieties make a colorful border for your garden, and their blossoms may be dried for flower arrangements. The white-flowered variety will grow better if you vary its location in your garden every couple of years. Yarrow may be started from seeds or root divisions and should be harvested when the flower is at its peak. Cut down the entire stalk, turn them upside down immediately, and hang them in your attic or in any dry, airy place out of the sun.

PUSSYTOES
Asteraceae
Antennaria neodioica

Pussytoes is a small, compact plant that has a cluster of basal leaves surrounding an upright, sparsely leaved stem. Atop this fuzzy stem sits a small cluster of tiny white flower heads, which, as you may have guessed, resemble the paw of a cat. The flowers, which are encased in the typical green bracts, appear from May to July. Pussytoes

PUSSYTOES
Antennaria neodioica

is commonly found in abandoned fields and old pastures in eastern Canada and the northeastern United States. The plant may reach eighteen inches in height in ideal situations.

Pussytoes has little use other than its ornamental appearance. An infusion made from the whole herb is known to stimulate the flow and secretion of bile and gastric fluids. Observing this plant is recommended over ingesting it, however.

MAYWEED, Stinking Chamomile
Asteraceae
Anthemis cotula

Yes, this common annual does, indeed, have an intense, fetid aroma. But, additionally, it has a powerfully acidic juice that is said to be capable of burning skin tissue at first touch. Nevertheless, a decoction and/or infusion of the herb has been used in American folk medicine as a remedy for cold symptoms, including congestion, fever, and headache. It is generally classified as a poison, however.

Mayweed is quite similar in appearance to its relative, Chamomile. It has the daisylike flowers, but the plant is somewhat more bushy than Chamomile. The leaves are finely veined and feathery. Mayweed may attain a height of about two feet under good conditions, and it flowers

MAYWEED
Anthemis cotula

throughout the summer months. It is an introduced species that is now found commonly throughout North America; it prefers disturbed areas, old fields, and roadsides for its habitat.

CHAMOMILE, Common Chamomile, Roman Chamomile, True Chamomile
Asteraceae
Anthemis nobilis

Chamomile is readily identified by its aroma, an applelike smell that is even more obvious when the plant is trodden upon. The herb is a creeping perennial, with hairy, branched stems and clusters of fuzzy leaves. The entire plant does not exceed a vertical height much over one foot, and its flowers, which bear a strong resemblance to those of the Daisy, appear from July to September. Chamomile prefers dry, sandy soil and plenty of sunlight.

The common name is said to come from two Greek words: *kamai,* meaning "on the ground," and *melon,* meaning "apple." The herb contains a bitter acid and therefore has a like taste, but it is used in the preparation of herb beers, virtually its only culinary use.

Medicinally, Chamomile has had a long history of applications.

CHAMOMILE
Anthemis nobilis

The herb has been cultivated extensively for its volatile oil, which is found principally in the flower heads. The famed Chamomile Tea, made from an infusion of the flowers, is said to be an effective sedative. It has been employed as a headache and nervous remedy, a digestive tonic, and an appetite stimulant. Chamomile is also said to be a powerful antiseptic and a practical application for abscesses.

Chamomile may be grown from seeds sown in late spring, but a better procedure is to use root runners to start the plants. Young seedlings should be transplanted to allow plenty of room for this sprawling herb. A sandy soil and plenty of sun are the only other requirements. The addition of Chamomile to your garden will generally improve the vitality of other nearby plants, according to old herbalists.

COMMON BURDOCK
Asteraceae
Arctium minus

It's hard to mistake this giant, sprawling herb. In the fall of the year, after its lavender flowers turn brown, the prickly little globes that are now the seed pods will hitch a ride on your clothes as you brush against the plant. On close inspection you'll find that, when flowering in July through October, the flowers of Common Burdock are com-

COMMON BURDOCK
Arctium minus

GREAT BURDOCK
Arctium lapa

prised of tubelike florets thrusting from green bracts with jagged-edged tips. The leaves can grow to a foot and a half in length, and are rather rounded, but pointed on the ends. Basal leaves are heart-shaped. The leaves are saturated with fine hairs underneath, and the leaf stalks are hollow. The veins of the leaves are easily visible. The herb grows to a height of five feet. There's usually no need to search for Common Burdock: it will find you as you travel through old fields, along road-sides, and in waste places across America and in Canada.

Common Burdock is useful in its entirety. Tender young leaves in the spring can be served as a green after boiling in two salted waters. The stems are also said to be edible, but they must be peeled first, or they will be much too pungent. They can be eaten raw, but many herbal adventurers say that boiling them is the only way to go. The flower stalk is said to be quite palatable when sliced and boiled much like the potato. But you have to harvest the foliage before the buds begin to take shape. Even the root, when properly peeled and simmered in baking soda water for twenty minutes, and again in plain boiling water, is said to be good to eat if gathered before the flowers appear.

Burdock was a favorite among Indians for curing many diseases. They thought it was a great tonic and blood purifier, and they used the herb as a diuretic. In the form of a salve, Common Burdock was used to treat burns and skin irritations. The root was the basis for many of the medicines from this plant. The leaves were supposed to aid indigestion and to tone the stomach. An infusion of the seeds was reputed to make a good wash for the skin, to smoothen it.

If you want to cultivate Common Burdock, make sure you have plenty of space in which to grow it. It's really quite big, and keep in mind that the roots can be up to a foot in length and an inch thick. In addition to starting from root cuttings, you may also sow seeds to begin your Burdock plants.

Believe it or not, there is an even larger species of Burdock, called Great Burdock (*Arctium lapa*), that will grow up to nine feet in height, or nearly twice the height of the Common Burdock. It is really an herb you can look up to, with its large flowering heads and grooved leafstalks.

SOUTHERNWOOD
Asteraceae
Artemisia abrotanum

A garden favorite, this shrubby perennial has a strong lemony scent when its gray-green, slender, hairlike leaves are bruised. Known as the herb of constancy, Southernwood grows to a height of about three feet and has a silvery appearance.

SOUTHERNWOOD
Artemisia abrotanum

The herb makes a nice dense garden border, where its lemony scent repels garden bugs. Branches of Southernwood, called "Aldermon" by early Dutch settlers, were strewn about the house to keep out the ants and the moths. Southernwood is used in some hair lotions, and it was formerly said to cure baldness. It is also used in baths and other aromatic liquids.

Southernwood could be brewed into a tea that was taken as an overall pep-me-up. A little honey was needed to offset its bitter qualities.

The herb found use as a stimulant and an antiseptic. Its aromatic flavor was used in Italian cakes. Southerwood was used to expel worms from children and to treat debility of the old.

This hardy ornamental is best propagated in the late summer by taking a six-inch-long cutting of the stem and burying it halfway into the sand until it roots. It needs a sunny, but sheltered, position with good draining soil, and it lends itself well to confined growing spaces.

WORMWOOD, Absinthe
Asteraceae
Artemisia absinthium

Wormwood, an ingredient in the infamous French liqueur, Absinthe, whose manufacture was banned in France in 1915 because it was thought to derange the mind, was nonetheless highly regarded for

WORMWOOD
Artemisia absinthium

its medicinal properties. The herb has a bushy, silvery appearance due to the presence of fine, downy hairs which cover the stem and leaves. The bottom leaves are large and deeply cut, and somewhat fernlike in appearance. But as one progresses up the multi-branched stem, the leaves become short, narrow, and blunt, and are interspersed with the tiny greenish-yellow globular flowers. The flowers appear from July to October. The fruits that follow do not have the characteristic hairs that most of the members of the Sunflower family have.

There are about 180 species in the *Artemisia* genus, including many common Western sagebrushes. They are all known to be extremely aromatic and bitter tasting, and with their somewhat narcotic properties, they have been known to cause dangerous irritation of the stomach, heart, and arteries. The liqueur made from Wormwood is often called "The Green Muse," because of its yellow-green coloration and its wicked effect on those who dared to imbibe. It is reputed to have the power, when taken to excess, of provoking hallucinations, acute mania, and general paralysis. The famous French painter, Henri de Toulouse-Lautrec, was said to have been fond of Absinthe; perhaps it contributed to his impressionistic creativity, but it is also said to have led to his demise. And with an alcohol content of 68%, it's no wonder. The liqueur was outlawed just before the First World War in France, the country of its origin, because officials thought it would lower the birthrate. Absinthe contained many herbal ingredients, including both Wormwood and Mugwort, Fennel, Balm Mint, Hyssop, and Star Anise, all in a brandy base. Anise was the dominant odor, giving the elixir a licorice aroma.

Chopped Wormwood leaves were said to be exceptionally good as a seasoning for roast goose, but their spicy bite was also used to flavor pork, beef, pickles, and salads.

Wormwood was used medicinally as a stomach remedy for indigestion, gastric pain, heartburn, and flatulence. It was thought to help relieve labor pains, and the oil acted as a local anesthetic when applied to rheumatic and arthritic joints. In addition, Wormwood was used to regulate menstrual flow, expel worms, and promote perspiration. It also had use as a sedative.

Wormwood is best planted in partial shade in well-drained soil. Raise it from seeds planted in the fall, or by root divisions set in early fall or spring. Stem cuttings set in early fall are also said to be effective. Make sure each plant has about two feet of surrounding space. The large plants should be staked, and in northern climes, it may be a good idea to cut back the plant and mulch it to protect it from freezing over the winter.

TARRAGON
Asteraceae
Artemisia drancunculus

Tarragon is one of the more popular herbs in use today. It is considered essential to French cooking. The herb is a perennial with a slender stem growing to a maximum height of three feet. Tarragon is unusual in that no seeds are produced following the white midsummer bloom. It must be propagated from root cuttings. The plant has smooth, dark green leaves.

The herb is occasionally called "Little Dragon," a rough translation of its Latin species name. It was said to cure the bites and stings of beasts and heinous creatures.

The oil of this widely used herb was used in perfumery. The leaves of Tarragon have been compared in flavor with Anise or Licorice. It is both sweet and slightly bitter. Fresh leaves of the herb are best, but if you must store them, keep them in a jar of vinegar. The vinegar will acquire the flavor of the herb. The leaves are indispensable in salads and sauces, like tartar sauce and Bernaise. Leaves and tops of the plant are both used in pickling mixtures, but Tarragon is probably at its best on fish, with roast meat and poultry dishes running a close second. It is also said to improve the taste of the relatively unflavored Artichoke.

TARRAGON
Artemisia drancunculus

A tea of the herb can stimulate the appetite, but at the same time help one to sleep when sipped just before bedtime.

When Tarragon was in popular use as a medicine, it was said to relieve digestive problems. It is also thought to be a diuretic and agent to promote menstruation.

Begin your plants with root divisions, and set them outside in March or April. They should be set a foot apart in well-drained soil with a lot of sun.

BIG SAGEBRUSH
Asteraceae
Artemisia tridentata

This cousin of Wormwood is abundant in Utah and other western states. The Indians employed it in a variety of tasks, from culinary to cosmetic. One of the most popular of the Sagebrushes, this herb grows to a height of ten feet. Big Sagebrush has a thick trunk and a few upward-reaching branches. The bark is gray and flaky, and the leaves remain green or yellowish-gray throughout the year. They possess a rich and biting aroma. Tiny yellow blossoms bloom late in the summer.

Early pioneers settled where the Big Sagebrush grew, and after a rain, the air was said to be filled with this herb's aroma. It can be

BIG SAGEBRUSH
Artemisia tridentata

burned in the fireplace to fill the house with a wonderful scent. When the fumes were inhaled, they were said to be a cure for colds and the flu. Indians would saturate their blankets in Sagebrush smoke, especially if there was an illness in the teepee, or after a childbirth.

Sagebrush shampoo was said to make hair grow, and a black hair dye could be made from it. The herb was said to make a good yellow dye, also. Indians burned Big Sagebrush to get rid of the odor resulting from an encounter with a skunk.

Medicinally, there were a number of uses for Big Sagebrush. Fresh or dried tea was used to stave off the flu, swelling, diarrhea, and indigestion. The herb was said to get rid of worms, sore throats, and sores. In the bath, it would supposedly get rid of aches.

You'll need plenty of sunshine and wide open space for this western heal-all. It grows best in dry, well-drained soil, from seeds or root cuttings.

MUGWORT
Asteraceae
Artemisia vulgaris

Mugwort and hot water make a refreshing bath in which to unwind and reflect on the day's activities. The perennial usually gets to be

MUGWORT
Artemisia vulgaris

about three or four feet tall, and its downy leaves and stem are covered with a soft, fuzzy hair. The color of the stems tends to be purple, while the leaves are a gray color underneath. Small globular flowers appear on thrusting stems. The flowers are yellow or reddish, and the seeds are gray. Mugwort is similar to Wormwood, but it has pointed leaves that are dark green on the top surface.

Mugwort is associated with St. John, and it is believed that, gathered and worn on St. John's Eve, which is similar to our Halloween, the wearer will be protected from any evil lurking about. John the Baptist is said to have clothed himself in this aromatic herb for protection in the wilderness.

One explanation of how this herb got its name is that it took the place of hops in the making of malt liquor. Mugwort tea is quite spicy, but it is said to have a number of medicinal properties. Mugwort is said to stimulate the appetite and regulate menstruation. In the bath, it is said to relieve the itching of Poison Oak, tired feet, and the pain of rheumatism. It has also been used as a purgative and digestive aid.

To grow some of this refresher in your backyard, choose a spot near the back of the garden so that it doesn't overtake your other plants. In better soil it will attain a substantial height. You can go to a nursery for seedlings, because you may not want to wait for the extremely long germination time of the seeds. Germination is more successful in temperatures in the mid to upper 50s.

MARIGOLD, Calendula, Pot Marigold
Asteraceae
Calendula officinalis

Brilliant orange and cheerful, the Marigold is, in addition, a versatile herb in the kitchen and medicine cabinet. The heavy stalk rises from a spring seed with resulting hairy leaves. The plant, with its distinctive flower, grows to a height of around eighteen inches. It is easy to grow and looks great almost anywhere, especially when planted in groups. Both the stalk and the clasping leaves are covered with fine hairs. Some of the leaves can get up to six inches long. The heavily scented flowers of this herb reappear almost as soon as one is clipped from the stalk. The plant puts on a late summer and early fall flowering drive.

The genus name is taken from the Latin, *calendae,* meaning "through the months." Marigold flowers open at dawn, as soon as the sun comes

MARIGOLD
Calendula officinalis

up, and fold in their golden petals as the sun sinks in the West. An old legend says that if you dream of Marigolds at night, it means that riches, success, or even marriage will come your way. Buddhists of India draped marigold flowers over themselves at ceremonial functions. And the Greeks made decorations to brighten up their homes from the orange blossoms. Conversely, the plant was thought by the Mexicans to signify death.

Marigold is known to be an inexpensive and readily available substitute for Saffron. Its gold color makes a delightful dye for foods, and the flower adds flavor to many dishes. Pound a few of the dried petals into butter, add some to the flour mixture of bread, or season stews and soups. The petals' flavor is brought out with cooking, and they can be included in rice, cheese, and egg recipes.

During the Civil War, the flower petals of Marigold were used to stop bleeding, and to this day, some herbalists utilize them to treat small wounds. Pills were made of the herb for the remedy of cancerous conditions. A salve made from the flower petals is said to give relief for sunburned skin. Marigold is also used as a skin softener. A pleasant tea can be made from the flowers which is said to perk up circulation and improve the complexion.

Choose a sunny location in your yard or garden for this fine herb. It is best grown from seeds in a rich, light soil. Early spring is the time to start, and the young seedlings should be spaced about a foot apart, so they can grow plump and round with numerous cuttings. Don't let the blossoms go to seed, or the plant will divert its energy from making more flowers. You may find that the addition of this plant to your garden will deter some common garden pests; the plant is often planted near the doors to houses. Harvesting the petals is fairly easy. When the flower is in full bloom, snip it off the stem. Line a screen with newspaper, and pluck the petals from the center of the blossom. Lay the petals on the paper so that they do not touch each other. Keep them out of the sun to dry, preferably in a cool, dark place. Since they must be turned often, a good way to do that is by using a second screen. Lay another newspaper on top of the drying petals, place the second screen on top of that, and flip it over. When the petals are crisp to the touch, jar them in an airtight container out of the sunlight. For winter blossoms, sow some of your Marigold seeds in midsummer, and bring them indoors just before the frost. You'll need supplemental lighting to insure at least twelve hours for each plant per day. The soil should be kept moist, but not soggy, for best results.

SAFFLOWER
Asteraceae
Carthamus tinctorius

A common substitute for the expensive Saffron, Safflower grows up to three feet in height and has a whitish, erect stem. The flower is

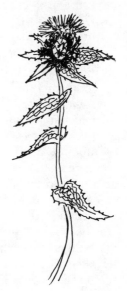

SAFFLOWER
Carthamus tinctorius

rather ball-shaped; it is fuzzy and yellow, until the approach of fall, when it slowly turns to orange. The leaves do not invite a touch, since they are covered with sharp, pointy spikes. The seeds ripen in August.

The flowers of this herb are prized for their petals, which contain an oil that mimics the oil of Saffron. No one knows where this plant originally came from, but it has two valuable dyeing properties, one producing a yellow color; the other, a red. The herb also makes pleasant shades of red when used to dye silks. When mixed with talc, the ground flower petals make an admirable rouge. The seeds contain an oil that is polyunsaturated, and so lends itself well to a salad oil for dieters.

The flowers of Safflower are also used as a laxative agent and for fevers in children.

Safflower grows anywhere but prefers a dry growing season in full sun. It is at its best in the West. Sow the seeds early in the spring where you want the flowers to grow, because the plants do not fare well when transplanted. Small rodents may wish to make a tasty meal of your Safflower, so protect it with a fence or a stealthy cat. Gather the seed head in the late fall, but be sure to wear gloves to protect your hands from the spiny leaves. Flower petals should be dried in late summer and stored in airtight containers for later use in cooking, dyeing, or cosmetics.

CORNFLOWER, Bachelor's Buttons
Asteraceae
Centaurea cyanus

Cornflower develops spectacular blue flower heads that seem to be positioned atop scaly oval bracts. The flowers bloom over the summer months, and they form at the top of long stalks. The plant grows to about three or four feet in height, and the grooved, branched stem is covered with a whitish, spider-weblike substance. Cornflower is found in waste areas and along roadsides in the northern United States and southern Canada.

According to an old legend, the centaur Chiron was once wounded by an arrow poisoned with the blood of the Hydra, and he bound his wound with the petals of the Cornflower, which counteracted the poison and healed the wound. Thus, the genus name of this plant was created, and the herb is said to have retained those healing powers to this day. Similarly, the species name, *cyanus,* is in honor of a legendary youth of the same name who was found dead in a field of his beloved Cornflowers. The goddess Flora is said to have transformed Cyanus into a

CORNFLOWER
Centaurea cyanus

Cornflower as a reward for his devotion to the plant. The herb has long been a symbol for delicacy as a result of the legend.

The flower petals of this herb have been employed medicinally in the past as a stimulating tonic and an eyewash of some repute. The juice of the petals is also said to make an impermanent blue dye. And of course, Cornflower is guaranteed by old herbalists to be an effective treatment of wounds and bruises, both internal and external.

SPOTTED KNAPWEED
Asteraceae
Centaurea maculosa

Spotted Knapweed is distinguishable by virtue of its spiny, pineapplelike flower bracts, upon which appears to be perched small, lavender flower heads. This herb grows to a height of about four feet, and its flowers usually bloom throughout the summer. It is commonly found along roadsides and in disturbed areas. It ranges from southern Canada to the northern half of the United States.

This member of the *Centaurea* genus has many of the characteristics of Cornflower. Its young stems and leaves are said to be edible either raw or cooked.

SPOTTED KNAPWEED
Centaurea maculosa

OXEYE DAISY
Chrysanthemum leucanthemum

OXEYE DAISY
Asteraceae
Chrysanthemum leucanthemum

Oxeye Daisy is said to be a woman's flower, because it has been found to be useful in the treatment of many ailments of the feminine body. The herb is quite common in fields, meadows, and along roadsides throughout most of the Continent. It is less common in the South. The white and gold flowers form on thin, upright stems; only one flower head appears per stem. The white portion of the two-inch flower head has all female rays, but the round, depressed yellow center, called a disk, is made up of both male and female flowers. The leaves are a dark green and are quite rough, coarsely toothed, and pinnately lobed. The basal ones get up to six inches long, and the upper ones attain only half that. Oxeye Daisy blooms throughout the summer season.

The genus name, *Chrysanthemum,* comes from two Greek words, *chrisos,* meaning "golden," and *anthos,* meaning "flower." The species name, *leucanthemum,* similarly, means "white flower." Farmers don't like to see this herb in their fields because it is said to give a bitter taste to the milk from cows that have ingested it.

The tender young leaves of Oxeye Daisy can be eaten in salads.

The leaves, stems, and flowers may be harvested in early summer and hung upside down to dry in a dark, airy place for future medicinal use. Oxeye Daisy makes a great tonic that helps in cases of bronchial trouble and asthma. It's especially good when drunk with a little honey. A poultice of the bruised leaves was said to reduce swellings and aid wounds, while a bath was thought to be good for palsy. The herb mixed with lard or petroleum jelly is reputed to be a good salve for wounds and sores.

CHICORY, Succory
Asteraceae
Cichorium intybus

From a long taproot, this perennial grows to a height of about three feet. It has a tough, twiglike stem with numerous branches. Leaves at the base of the plant are large and fuzzy, resembling Dandelion leaves, but those further up the stem are small and stalkless. The flowers, which bloom from early summer to late fall, are light blue in color. They form in the leaf axils as a large cluster and, interestingly, the flowers are known to open and close at specific times every day, depending on the latitude of residence of the plant.

Chicory is known by many as a coffee substitute or coffee additive,

CHICORY
Cichorium intybus

and it is a close relative of the common Garden Endive, a popular salad green. Chicory was said to have been eaten as a potherb itself by the Romans, and, indeed, is still used as such today. It is also commonly grown for animal fodder. The plant is heavily cultivated in Europe for its root, which, when ground and roasted, provides the popular coffee alternative. Its addition to coffee mixtures is said to produce a slightly bitter taste and a darkened color, but the herb lacks most of coffee's annoying side effects.

The root of Chicory is also said to contain some sedative properties, and a decoction of the same is said to have been used to treat certain internal complaints, including jaundice, gout, and rheumatism. A children's laxative has been concocted from the herb, also. Old herbalists warned, however, that excessive use of this herb might lead to intestinal congestion and possible loss of vision.

Chicory is common in the wild all over the world, but it is also cultivated from seeds sown in late spring. Simply allow it enough room to grow, water it in dry weather, and watch it grow.

BLESSED THISTLE, St. Benedict Thistle
Asteraceae
Cnicus benedictus

Blessed Thistle is an annual sporting a raggy yellow flower; it grows to a height of two feet. It is used nowadays mostly as an ornamental, but was once regarded as a heal-all in the past. It has long, silvery green leaves that are lance-shaped. Both the stem and the leaves of this coarse plant are covered with hair. The flowers appear from May to August and are nestled amidst the upper leaves at the end of the branches. A bristly yellow bract encases the flower head.

Blessed Thistle is a native of southern Europe, and it grows commonly in waste areas. It was thought to counteract the effects of various poisonous substances, especially those affecting the heart.

The herb was said to be an overall tonic, that would make one perspire, thus ridding the body of harmful agents. It was reputed to be a cure for the plague, and it was thought to be effective to rid one of worms. Blessed Thistle was used as a general purgative for the entire system. A tea of the herb was used to fight fever.

Sow the seeds of Blessed Thistle in the spring of the year in the best soil, or the worst, but give it plenty of sunlight. Don't forget the water.

BLESSED THISTLE
Cnicus benedictus

JERUSALEM ARTICHOKE
Asteraceae
Helianthus tuberosus

This variation of the Common Sunflower was once a valuable food source for the American Indians, who cultivated it for its tasty tuberous roots, which are said to have a sweet, nutty flavor. Jerusalem Artichoke grows to a height of ten feet. Its thick, rugged stem contains some branches, as well as the tough, oval-shaped leaves that attain a ten-inch length. Atop this tall plant sits the well-known yellow flower head; it appears from late summer to early fall. Jerusalem Artichoke prefers a moist soil, and it has spread from its early cultivated areas to its present range, which includes most of North America.

The common name of this herb comes from a bastardization of the Italian word, *girasole,* which means "turning to the sun." The flower of this herb is quite attractive, but it is also known to have properties enabling it to curdle milk.

Lewis and Clark were said to have dined on the tubers of this plant, a culinary custom taught to them by an Indian squaw. Previously, the tubers have been baked into pies with raisins, dates, and spices. They were frequently boiled until soft. After peeling, they could be sliced and placed in stews with butter, wine, and spices. The flavor of

JERUSALEM ARTICHOKE
Helianthus tuberosus

Jerusalem Artichoke tubers is not agreeable to everyone: some have likened it to what one might imagine soft ash from a fireplace would taste like. The tuber is said to be highly nutritious, however, while lacking in the starch content found in the potato.

For best results, Jerusalem Artichoke should be cultivated in a light, rich soil. The plants may be started from the tubers or parts thereof, set in the soil at the end of winter. They should be placed six inches under the surface, and plenty of room should be allowed for this rather large plant's growth. As you weed in the early summer, push some soil up around the stem base to help anchor the plant. Like potatoes, you should not dig up the tubers until the plant's leaves have died down entirely, insuring maximum tuber growth. After the tubers are dug, they must be kept from light and air, or they will turn dark when cooked. A good way to do this is to keep a box of sand or dirt in a cool area of the house in which to nestle the tubers so that all parts are covered.

ELECAMPANE
Asteraceae
Inula helenium

This herb has a thick, perennial root shaped like an inverted cone, from which grows a six-foot-tall hairy stem. Long, basal leaves emanate from the stem on long stalks and may sometimes attain a length of

ELECAMPANE
Inula helenium

twenty inches. Upper stem leaves are large, green, and oval-shaped, with hairy undersides. The leaves are generally coarse and toothed. The flower of Elecampane, which appears from July to September, is a large, yellow, Sunflowerlike head attaining four inches in width. The plant prefers the dry, well-drained soil of old fields and roadsides, and it may be found growing wild in southern Canada and, in general, most of the northern United States.

The root of this giant herb was most widely used in the past, for both its culinary and medicinal value. In Europe, a candied confection was made of the root, which in addition to its sweet tastiness, was said to be a good overall aid to digestion and asthmatic complaints. The active principle contained in the Elecampane root is Inulin, which is principally used in congestion and pulmonary complaints. It is also said to be somewhat astringent and to possess marked antiseptic properties.

Elecampane is easily cultivated from root divisions set in the fall. Make sure each division has a bud. Allow plenty of room for this giant, and pick a somewhat shady location in a fairly good, well-drained soil.

SWEET GOLDENROD, Blue Mountain Tea
Asteraceae
Solidago odora

The bruised foliage of Sweet Goldenrod has a Licoricelike aroma that facilitates its identification in the wild. It is a relatively tall plant,

SWEET GOLDENROD
Solidago odora

growing to a height of around three feet in fields and along the edges of woodland in most of the eastern United States. The long, narrow leaves of Sweet Goldenrod grow right from the erect stem and have no stalks. The flowers form bright yellow clusters on drooping branches throughout the summer months.

As its alternate common name implies, Sweet Goldenrod has been used by both the American Indians and the Pennsylvania Dutch to make a sweet tea, which is said to dispel flatulence, act as a diuretic, and make a good stimulative. The Indians reportedly made a lotion for bee stings from the flowers of this herb, also.

SLENDER FRAGRANT GOLDENROD
Asteraceae
Solidago tenuifolia

This herb differs from its cousin, Sweet Goldenrod, in that is has distinctive flat-topped flower clusters, which, nonetheless, are still quite yellow. The stem and leaves are quite similar, but this species may grow a little taller, to about four feet in height. One slight difference exists in the leaves of Slender Fragrant Goldenrod: they are somewhat grasslike and have only one rib. The plant is found in the same habitat as Sweet Goldenrod, but its range is slightly extended to the north into Canada, and to the west into the Dakotas.

SLENDER FRAGRANT
GOLDENROD
Solidago tenuifolia

A tea made from the leaves of this herb is said to stimulate kidney action, control inflammation, and heal wounds. Additionally, the crushed leaves are said to make a good antiseptic powder.

COMMON TANSY
Asteraceae
Tancetum vulgare

Feathery leaves and a heavy scented flock of yellow flower heads mark this plant. Common Tansy grows to a height of three feet. It is indigenous to Europe, but has been widely cultivated in this country from the days of the Revolutionary War. The leaves have a bitter taste, grow up to five inches long, and are deeply cut. The creamy yellow flowers have no petals, and they remain on the plant for a long time, appearing in midsummer and lasting into the fall. The seeds that follow remain on the plant throughout the winter and drop off to root in the spring.

Common Tansy flowers were often strewn about to keep flies away. The flowers have an aromatic, bitter taste, and have been used to expel worms and treat hysteria. The herb was at one time used to preserve the bodies of the dead. A cake of the herb combined with eggs, cream, Spinach juice, and spices was served at Lent.

COMMON TANSY
Tancetum vulgare

Grow this herb near the walls of your house, and it is said that the ants won't come in. Otherwise, the herb can be grown in your garden from seeds, in average, well-drained soil. Common Tansy plants can also be started with root divisions.

COMMON DANDELION
Asteraceae
Taraxacum officinale

Often unappreciated by suburbanites, and the object of derision in television commercials, Common Dandelion is, on closer inspection, sometimes found to be attractive and even useful. Solitary yellow flower heads are perched atop slender, foot-long stalks that emanate from long basal leaves. The leaves are toothed and lobed, and the fruit is comprised of a single seed for each of the tiny flowers in the overall flower head. Each seed is attached to white, silky hairs forming a globelike, silver ball which replaces the flower head.

Children often unwittingly help to propagate the Dandelion by making a wish, blowing on the silvery seed head, and scattering the tiny seeds into the wind. Common Dandelion may draw its name from a comparison of its leaves with the teeth of a lion. The herb is thought by some to have been introduced from Europe.

The leaves of Common Dandelion can be gathered as a salad or potherb. They also make an interesting green wine. The roots have been roasted as a coffee substitute like the blue-flowered Chicory. Indians chewed the stems of this herb much as we do chewing gum. Common Dandelion tea is said to be a good iron tonic, and may revive a slow-acting liver.

Common Dandelion roots and leaves are recommended alone or in combination with Carrots to form a fortifying juice. The herb is rich in vitamins and minerals. It is said to normalize the acid content of the body. It can be made into a soothing bath for stiff joints, or a skin toner. Dandelion is said to aid digestion, quell fevers, and give restful sleep to the fitful.

If by chance you don't already have enough of this herb in your yard or your neighbor's, capture a few of the parachuting seeds and place them where you want them, under ¼ inch of soil. A faster way to start them would be to transplant a portion of the long taproot. That same quickly-regenerating taproot is the cause of difficulty in eradicating this plant from your lawn.

All parts of the plant can be eaten; the greens mixed with a little vinegar make a good blood purifier, according to herbalists. But pick the greens before the blossoms appear. The juice of the plant is said to remove warts; furthermore this liquid substance, like latex, may be a potential source of rubber compounds. The roots of Common Dandelions make a magenta dye.

COMMON DANDELION
Taraxacum officinale

COLTSFOOT, Coughwort
Asteraceae
Tussilago farfara

Coltsfoot has separate stems for flowers and leaves, and the former actually appear before the latter, a curious trait that led the herb to be known by an old Latin common name which, roughly translated, meant "the son before the father." A small plant, Coltsfoot resembles Dandelion. It has a single yellow flower head, and the leaves are, of course, shaped like a colt's foot. The flowers blossom from late winter to early summer, and this common herb can be found in waste areas from Canada to the northeastern United States.

Coltsfoot has a great reputation in the herb world as a cough remedy. Its generic name, roughly translated, means "cough dispeller," and much reference to that property has been made in various ancient herbals. A tobacco substitute is made from the leaves and smoked to effect the cure. Additionally, a decoction of the leaves is often mixed with Licorice or honey to make a cough syrup, cough drops, and hard candy. A tea is also made from the leaves, and an extremely thick decoction of the herb has been used in the past to treat skin afflictions.

Coltsfoot is quite common, but should one desire to cultivate it, it can be grown from seeds or stem cuttings, the latter being the more effective method.

COLTSFOOT
Tussilago farfara

BALSAMINACEAE
The Touch-Me-Not Family

Mostly herbs with succulent stems, the members of this family have alternate leaves with featherlike veins. The flowers have two lips and are rather showy. One sepal has a large, curved spur; two others are small and greenish. Yet two others are very small or missing altogether. There are five stamens which are often united. The fruit is oblong and forms as a capsule containing many seeds. Botanists say 2 genera and 400 species exist, mostly found in the tropics of Africa and Asia, but with some examples found in North America as well.

JEWELWEED, Spotted Touch-Me-Not, Glassweed
Balsaminaceae
Impatiens capensis

Jewelweed will produce a warm, golden color when used as a dye for wool. The plant is tall, growing to a maximum height of five feet, and it is particularly fond of swampy, wet areas. The stems are succulent and translucent. The spurred flower is an attractive golden orange,

JEWELWEED
Impatiens capensis

with small reddish dots, as though someone had sprinkled it with pepper. The leaves are rather long and oval, and they have a pale green surface on top, but underneath they are grayish or blue. The fruit is rather curious, and the herb derives its name from the fruit's action when it becomes mature. Even the slightest touch of the capsule will cause it to burst open, spewing its seeds everywhere. Jewelweed flowers from July to October and is found in most of North America.

The other common name, Glassweed, is based on a reference to the plant's brittle stem, which will break like glass when touched. Jewelweed is said to symbolize "impatient resolve."

One herbalist says that the stems of Jewelweed are quite delicious when cooked or raw. The herb is valued for its high moisture content. Other herbalists, however, have warned that the plant is poisonous when taken internally and may cause mild vomiting.

Jewelweed has been used to stop the spread of Poison Ivy. The usual procedure is to break off a succulent twig of Jewelweed and smear the juice on the affected part. It is sometimes recommended that one smear himself with the juice before venturing into the woods, if a susceptibility to the itchy rash is known. The herb has also been used as a treatment for athlete's foot, because of its supposed antifungal properties.

Jewelweed will grow from seed in the spring. It grows best in shady, wet areas. It is a great attraction for hummingbirds and honey-making bees.

BERBERIDACEAE
The Barberry Family

The herbs of this family have either simple or compound leaves and clustered flowers. Their fruits are usually berries. Botanists say there are 9 genera and about 600 species, a few of which are cultivated as ornamentals.

MAYAPPLE, Mandrake, Duck's Foot
Berberidaceae
Podophyllum peltatum

This early spring plant eventually produces a single white flower which emanates from the crotch between the herb's two large, lobed leaves. It commonly blankets the forest in the spring, and Mayapple develops a large pulpy lemon-shaped berry around June or July. The herb is quite common in shady woods throughout most of Canada and the United States.

Mayapple derives one of its alternate common names from two Latin words meaning "foot" and "leaf." They are the source, addi-

MAYAPPLE
Podophyllum peltatum

tionally, of the genus name, *Podophyllum*. The American Indians were said to have used the root of the herb as a cathartic. In quantity, the root is said to be poisonous and has been used to commit suicide.

The mature fruit of Mayapple is said to be safe to use in jellies, or it can be eaten raw. The juice can be made into a lemonade-type drink. The seeds should not be ingested.

The herb has held some repute as a nonaddictive laxative and as an agent used to promote the flow of bile. Large doses will irritate the digestive system, however. Mayapple's active principle, Podophyllin, has been experimented with recently in the treatment of paralysis and as a professed cancer cure.

BORAGINACEAE
The Borage Family

The herbs of the Borage family are often velvety or fuzzy in appearance. Their flowers are symmetrical, five-petalled, and often found in clusters. The leaves are simple in structure. The fruit is sometimes a berry, but most often a four-sectioned nutlet. Within the approximate 100 genera, there are about 2,000 species. Members of the Boraginaceae family are found mainly in the temperate regions.

ALKANET, Dyer's Bugloss
Boraginaceae
Anchusa officinalis

Alkanet grows to a height of about two feet under ideal conditions. It is a biennial plant, and its violet flowers are found in small clusters that bloom from July to October. The leaves are narrow, hairy, and oblong; they are from three to six inches long and about an inch wide. They are dark green in color and rough in texture. Alkanet favors southern climates, but it does extend north to Canada along both coasts of the United States.

ALKANET
Anchusa officinalis

The genus name *Anchusa* is derived from the Greek *anchousa*, which means "paint." The root has long been used as a source for a red dye. The dye, which is extracted from the outer covering of the root, was once used to color salves and to stain wood. The color lends itself readily to oils and alcohol; it turns water brown. French women were said to have brightened their faces with a red hue created by the herb.

Greek herbalists said that a decoction of wine and Alkanet would strengthen the back and ease pains. English herbalists said, moreover, that a vinegar made with the herb would remedy yellow jaundice and aid in problems with the spleen and kidneys. The whole herb is used medicinally by American herbalists to expel worms and prevent fluid retention.

The purple flowers can revive a glass of wine with a fresh Cucumberlike taste and brighten a fruit or vegetable salad uniquely.

Alkanet thrives in full sun, but it will do well in partial shade. Use plenty of compost on a well-drained soil when you start your seedlings in the spring. But when starting from seeds, the plants will take longer to mature than by dividing roots and setting them in rows about a foot apart. The seeds like about ¼ inch of soil to cover them while germinating. There is yet another way to begin your dye plants: put two inches of soil over potted root cuttings in the fall of the year. Be sure the roots are upright for the most efficient growth pattern. The container can then be set aside in a cool basement until the threat of frost is gone from the spring.

BORAGE
Boraginaceae
Borago officinalis

Borage stands about 1½ feet high, and its rough, hairy stem has numerous branches. The herb is a succulent, and the stem is hollow. The leaves are somewhat wrinkled, thick, and oval-shaped. The flowers are five-petaled and blue, and they have distinctive black anthers. The fruit is the typical dark brown four-sectioned nutlet. Borage is considered a garden escape and may be found growing on rubbish heaps or in waste places. It is usually best grown for kitchen use, however.

Borage has long been thought to symbolize bluntness or rudeness, supposedly because it was thought also to have intoxicating properties. For that reason, it was commonly found as an ingredient in many drinks. Some have called Borage the "most attractive of all herbs,"

BORAGE
Borago officinalis

and it has, for that reason, often gone by the common name "Herb of Gladness."

A tea made from the leaves of Borage is said to give a feeling of exhilaration. The herb has been used as a laxative, an eyewash, and an agent to induce sweating. As a face pack, Borage leaves are reputed to relieve dry skin. The herb is also said to increase the resistance to insects of plants growing near to it.

The young tops of Borage were often eaten as a pot herb and were also added to soups for flavoring. The leaves will give a Cucumber flavor to pickles, and they have been boiled, chopped, and served with butter like Spinach. The flowers are sometimes candied or crystallized for confectionery use.

Borage is best grown from seeds; it does not transplant very well. It has a rather long tap root, so it isn't a good bet for indoor growing. Outdoors, though, any light, well-drained soil in a sunny location will suit Borage. Just give it about two feet of space on all sides, and it will be happy.

FORGET-ME-NOT
Boraginaceae
Myosotis scorpioides

Forget-Me-Not is a hairy, spreading plant that grows to a height

FORGET-ME-NOT
Myosotis scorpioides

of two feet. The small leaves are a blunt oval shape and, for the most part, have no stalks. The five-petalled blue flowers have a distinctive yellow "eye," and seem to sprout from a tubular base. They grow on the many small branches that unfurl as more flowers come into bloom. The flowers appear throughout most of the summer, from May to October.

Forget-Me-Not was a symbol of remembrance in Medieval times. The species name, *scorpioides,* refers to the shape of the unblossomed flower cluster. This was a big hit with the European settlers, who brought the plant along with them to the New World to plant in their personal gardens, perhaps as a remembrance of the loved ones they left behind. Escapees from those early gardens have now found their way to the shores of lakes and streams in most parts of the continent.

COMFREY, Boneset, Knitbone, Bruisewort
Boraginaceae
Symphytum officinalis

Comfrey is a common plant in wet pastures and meadows every- where. From a rootstock that contains a thick juice grows a bristly stem with pointy oblong leaves. The leaves, also, are somewhat hairy, and the basal leaves of the herb are a little rounder. The flowers, which appear from May to August, are a pale purple color. They are found in pairs and form a curve like the tail of a scorpion, with the blossoms getting smaller at the end of the curve. Comfrey grows to a height of about three feet.

Comfrey is famed for its supposed ability to speed the healing of broken bones. Its common name is said to be a corruption of the Latin *con firma:* a reference to the above trait. Also, the genus name, *Symphytum,* comes from the Greek *symphyo,* which means "to unite."

Comfrey root is high in mucilage, and it is, therefore, used somewhat like Marshmallow. It is said to be good for digestive disturbances and intestinal troubles. It is commonly applied as a poultice, also, and is said to have the ability to reduce swelling and pain in inflamed areas. Speculation has it that Comfrey's tendency to diminish swollen parts is the trait that gave it its bone-setting reputation, since any reduction of swollen tissue surrounding a break will aid in the bone's reunification.

Comfrey is said to be high in calcium, potassium, phosphorus, and other trace mineral content. Additionally, it is said to be the only land plant to contain vitamin B-12, which is most often found in animal sources. It is also a good source of lyscine, an amino acid that many strict vegetarians lack. A good coffee substitute has been made from a mixture of Comfrey root, Dandelion, and Chicory.

Comfrey can be grown almost anywhere in partial shade. It can be started from seeds, or by root cuttings sown in the fall. Space the plants at about two and one-half feet apart.

COMFREY
Symphytum officinalis

BRASSICACEAE, or CRUCIFERAE
The Mustard Family

The herbs included in this family are distinguished in part by their tangy-tasting sap, or mustard-flavored rootstock. Many of the members of the family are well-known garden vegetables, such as Cabbage, Cauliflower, Broccoli, and Brussel Sprouts. Plants are usually small, with leaves that are either simple in structure or pinnately divided. The flowers are small, with four petals, usually six stamens, and four sepals. All the flower parts are attached near the base of a two-chambered ovary. The fruit is a pod. There are said to be about 375 genera and approximately 3,200 species.

COMMON WINTER CRESS, Common Mustard Hedge, St. Barbara's Hedge
Brassicaceae
Barbarea vulgaris

The bright yellow flowers of Common Winter Cress are a welcome sight in moist fields, meadowlands, and brooksides from the early spring

COMMON WINTER CRESS
Barbarea vulgaris

through August. The flowers appear in elongated clusters atop an erect leafy stem. The tiny cheery flowers, like other members of the Mustard family, form a cross with their four little petals. To identify this herb, look for lower leaves up to five inches long that are pinnately divided into five segments. The segment in the middle is large and rounded, but the leaves at the top of the plant are considerably different. They are lobed and appear to partially surround the stem. The fruit is an erect seedpod that has a short beak. Common Winter Cress is found in the wild in Canada south to Virginia and west through the corn belt. It usually grows from one to two feet in height; it is an introduced species from Europe.

Common Winter Cress makes a tasty and colorful addition to a spring salad. Sometimes called Yellow Rocket or Upland Cress, *Barbarea vulgaris* can be cooked and eaten much like its cousin, Spinach. The bottom leaves may have a somewhat bitter taste, but they may be substituted for another cousin, Watercress.

The hardy biennial, with its green and glossy leaves, likes a rich and moist soil. Seeds can be sown in the late fall, or allow the fruit pods to burst and spill their seeds to regerminate the fertile soil of your garden. Common Winter Cress can also be grown from root cuttings gathered in the spring.

WHITE MUSTARD
Brassicaceae
Brassica alba

White Mustard is closely related in appearance to Black Mustard, but the plant is somewhat smaller. In addition, the seedpods of White Mustard extend horizontally from the stem, while those of Black Mustard are smooth and erect. The seedpods of White Mustard are also somewhat fuzzy or hairy. Each seedpod contains about four to six round seeds which are yellow in color both internally and externally. As with Black Mustard, White Mustard is commonly found in fields and along roadsides or waste areas. The plant is smaller in size than Black Mustard, growing to only a foot in height. The flowers are somewhat larger, but are still yellow.

The inner seedcoats of White Mustard contain some mucilage which enables them to be utilized as a moisture absorbent, especially in chemists' bottles. Both Mustards, but especially the White, are used as fodder for sheep, and as a cover crop to prevent the loss of nitrates in the soil.

WHITE MUSTARD
Brassica alba

White Mustard is said to make an excellent salad green. At least one old herbalist is said to have preferred the herb to lift the spirits and refresh the memory.

The seeds are usually ground into a powder which is somewhat less potent than that of Black Mustard. The powder is usually mixed with the black variety to make Mustard flour. White Mustard seeds were once used as a laxative, often with less-than-pleasant results. A gargle made from an infusion of the seeds has been recommended for sore throat.

White Mustard is commonly raised in gardens, and it will grow year-round indoors. The seeds germinate very quickly.

BLACK MUSTARD
Brassicaceae
Brassica nigra

Black Mustard is quite common in open fields and waste areas throughout most of the United States and Canada. It grows to a height of about three feet and is noticeable by virtue of its many branches. The plant has tiny flowers that are yellow in color and which appear in small clusters near the top of the stem. The flowers appear from June through October. The leaves of the herb are lobed, with one main lobe about three inches long, accompanied by four lobes on the sides of the leaf. The upper leaves of the plant are simply toothed. The fruit is a pod that grows close to the stem. Each pod contains about ten or twelve reddish-brown or black seeds.

While the ancient Greeks held Black Mustard in high esteem for its medicinal properties, the Romans more commonly made use of the herb for culinary purposes, both as a green vegetable and a condiment. Mustard was formed into balls with honey or vinegar, originally, and the balls were simply mixed with more vinegar when the mustard was desired. Then, supposedly in the eighteenth century, an English woman discovered a method to make Mustard flour, which made long-term storage a lot easier.

Black Mustard has long been employed as a condiment to accompany meats of all kinds. Its leaves can be eaten as a salad green, but they are quite pungent. The seeds are sometimes used as a seasoning in pickle recipes, and are said to be a favorite food of songbirds, for which they are commercially prepared.

Black Mustard is commonly used medicinally as a poultice applied to the area of internal inflammation. It is said to relieve congestion of various organs by drawing the blood to the surface of the skin. The oil contained in ripe Mustard seeds is a powerful skin irritant, but when prepared in combination with other compounds, it produces an excellent liniment, supposedly. Throughout history Black Mustard seeds, flour, and oil extract have been used variously for making a footbath to dispel colds, curing hiccoughs, growing hair, making an oil used in precision clockworks, treating epilepsy, and clearing the voice.

Mustard is so common in the wild that one would have little reason to grow it indoors or in a garden. It can be done, however. Seeds are best sown in the spring in rich soil about a foot apart. The plant does not like excessive moisture. The valuable seeds ripen at the end of summer, and they are usually husked and ground into flour.

SHEPHERD'S PURSE
Brassicaceae
Capsella bursa-pastoris

The unique heart-shaped seedpods of this common herb resemble to such an extent the purse once carried by shepherds in the Middle Ages, that the common and Latin names assigned to the herb are a direct reference to that article. Shepherd's Purse is an erect annual sporting a terminal cluster of tiny white flowers. The petals are arranged in a cross formation. The basal leaves are deeply toothed and resemble Dandelion leaves. The stem leaves are smaller and are shaped like clasping arrowheads. The fruit is a triangular, wedge-shaped pod that looks like a tiny pillow which is indented at the tip. The plant grows to a height of about one and one-half to two feet in disturbed areas, lawns, and roadsides. It is likely to turn up anywhere in southern Canada and the United States. It flowers from March to September.

Shepherd's Purse was widely used as a food plant and as a medicinal plant in earlier times. Small birds love to eat the seeds of this herb. But when consumed in large quantities by cows, Shepherd's Purse taints the milk with its peppery flavor. In World War I, when clotting agents were in short supply for both soldiers and civilians, a liquid extract made from the herb was used for that purpose.

Shepherd's Purse contains vitamins C and K, and the minerals calcium and sodium. The leaves have a peppery taste and can be eaten raw or cooked. Revolutionary War soldiers at Valley Forge were said to have relied on Shepherd's Purse for sustenance during the cold winter. The American Indians were fond of using the seeds as a food source. The seedpods, when dried, make a pepperlike seasoning. Toss some of the leaves with a little oil and vinegar for a tasty salad, or blanch the leaves and eat them as you would other greens. A stimulating tea made from the leaves is said to be quite aromatic.

The same tea is said to raise blood pressure and is active in a number of other medicinal ways. Shepherd's Purse tea is said to have

SHEPHERD'S PURSE
Capsella bursa-pastoris

a stimulating effect on the kidneys. It supposedly also helps to stop internal hemorrhaging in the kidneys, lungs, stomach, and uterus. The herb, when applied to cuts or wounds, was said to help the healing process. It was also used to stop nosebleeds, earaches, and irregular menstrual cycles.

Shepherd's Purse can be grown from seed in the early spring in any kind of soil anywhere. But it will grow larger in better soil. Place the seedlings six inches to a foot apart to make room for the branching stems. If left to go to seed, Shepherd's Purse will propagate itself with ease.

SPRING CRESS, Bitter Cress
Brassicaceae
Cardamine bulbosa

Spring Cress has small white flowers with the four petal arrangement in the shape of a cross. Flowers appear atop a smooth erect stem. They are commonly found in numerous clusters on the two-foot-high stem. The leaves that form at the bottom of the stem have long stalks, are oval-shaped, and up to an inch and a half long. Along the stem, the leaves are lance-shaped, usually toothed, and without stalks. The seedpod forms an erect, narrow fruit. Spring Cress grows along springs, swamps, and wet clearings, favoring the latter two. It can be found in the Midwest from Texas to Minnesota, and in the East from Canada to Florida.

Spring Cress is popularly collected in the spring, just after the last vestiges of winter melt away. Its leaves and stems taste just like mus-

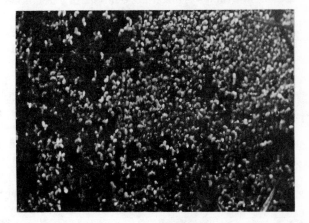

SPRING CRESS
Cardamine bulbosa

tard, but most people don't associate it with its cousins in the Mustard family. It is used in salads, soups, and sandwiches.

Spring Cress can be propagated along streams, wet clearings, and swamps. Just watch the plants take root!

PENNSYLVANIA BITTERCRESS
Brassicaceae
Cardamine pensylvanica

Pennsylvania has its own native variety of Watercress in the form of this nonaquatic species. Pennsylvania Bittercress, which grows along the banks of streams, but not in them, can be distinguished from its water-loving cousins by the little teeth on its many small leaves. It is also found in Ohio, New York, and New Jersey.

The greens can be used much like the other Brassicaceae members in salads, soups, and sandwiches.

WOAD, Dyer's Weed
Brassicaceae
Isatis tinctoria

Woad is a yellow-flowered biennial that has been extensively cultivated for its use as a dye. In recent history it has been left behind in favor of indigo, but it is still used as an ingredient in indigo dye, and it is said to greatly improve the color and quality of that dye. In its second year of growth, Woad comes up with a one and one-half to four foot stem which has alternate oval-shaped or lance-shaped leaves which actually grow almost completely around the stem at their bases. The bottom leaves are about four inches long, while the upper ones decrease in size to about one and one-half inches at the top of the stem. The leaves are blue-green in color. The flowers, which appear from June through September, are found in multiple clusters that may branch out as much as two feet in width. When the flowers appear, the black, tonguelike seeds follow soon thereafter.

Today, Woad is more commonly used in flower arrangements and potpourris that it is in the making of dyes. An old historical reference states that British females used to stain their nude bodies with the herb and then attend sacrificial rites.

Woad is said to have a sharp taste and is not recommended as a medicine to be taken internally. It at one time was said to be useful

WOAD
Isatis tinctoria

externally when made into a plaster form to relieve problems with the spleen and skin ulcers. Woad was also said to be good at controlling inflammation and bleeding.

Woad is a hardy plant that grows anywhere from southern Canada to the Gulf of Mexico. It prefers a loamy, well-drained soil in full or partial sun. Seeds are best sown in rows from six inches to one foot apart, preferably in late summer. Seedlings can be transplanted when they are about three inches high. Spring is a good time for transplanting when the ground is moist. If Woad is allowed to go to seed, you'll be sure to have more Woad plants the following year. You may find a market for your Woad among the local weavers' guilds, whose members may be seeking its unique traditional color for their wool. Make sure you don't grow Woad in the same place for more than two years in a row, because it depletes certain elements in the soil. The leaves should be harvested when they are mature (fully grown) and green. The highest quality of dye is obtained from Woad just before it comes into flower. If you're lucky, you'll get three crops a year.

GARDEN CRESS, Pepper Cress
Brassicaceae
Lepidium sativum

Garden Cress is an annual that grows to about one and one-half feet in height. It is another spring plant with a peppery flavor that is used as a salad green. The tiny white flowers that form at the top of the stem have four petals that form the shape of a cross, a common trait of the Brassicaceae family. The bottom leaves are deeply lobed and have long stems. Garden Cress is common in southern Canada and most of the United States.

Garden Cress is definitely a kitchen plant. Its peppery flavor, found in both the leaves and stalks, can be used as a base for soups and makes a smashing sandwich when spread over cream cheese or mixed with its cousin, Mustard. Garden Cress tastes best when the plant is very young. It will pep up a dull salad with a refreshing new taste.

To grow Garden Cress use the seeds, since the plants you use may not have time to mature and propagate themselves naturally. Pick a sunny spot outdoors with a rich, but well-drained soil; or prepare window boxes for indoor germination. Spread the seeds evenly in your planting space. Press them into the soil, water them, and place in a brightly sunlit area. Garden Cress likes a cool germinating temperature of not less than 65 degrees Fahrenheit. Some herb fanciers like to grow Mustard and Garden Cress together for convenience. Keep in mind the Cress seeds will take longer to germinate than the Mustard, so it's best to give your Garden Cress a four-day advance planting.

GARDEN CRESS
Lepidium sativum

TRUE WATERCRESS
Nasturtium officinale

TRUE WATERCRESS
Brassicaceae
Nasturtium officinale

True Watercress is an aquatic plant whose stems float on the surface of the water. The fleshy stem first creeps over the mud before shooting up to a foot in height. The peppery tasting leaves of this perennial are shiny and somewhat rounded. The leaves range from one to six inches long and have five to nine rounded parts. The terminal segment stretches above the side parts. The fruit appears in the form of an inch-long, upward-curving seedpod. Small white flowers appear on the plant in springs, brooks, and cool streams from April to October. True Watercress is found everywhere from southern Canada to the Gulf of Mexico.

True Watercress is grown commercially in Pennsylvania, Alabama, Virginia, Florida, and California. It originally came from Europe, however. The herb is especially prized in New York and Chicago for its zesty flavor in the making of sandwiches, salads, and as a garnish in meat dishes.

Both the leaves and stems are edible. Watercress was often called Scurvy Grass because it is rich in vitamin C and iron and was used from the earliest times to ward off that nutritional disorder. In addition the plant also is said to contain vitamins A, D, and E, along with the minerals Fluoride, Iodine, Iron, Magnesium, Nitrogen, Niacin, Phosphorus, Potassium, Sodium, and Sulphur. True Watercress can be used as a base for soup which is properly served either hot or chilled. It's best to eat fresh Watercress, but it can be frozen for later use.

An infusion of the leaves of True Watercress is said to relieve indigestion, rheumatic pains, and bronchial congestion. Juice from the leaves and stems has been used to take away skin blemishes. Herbalists say that the juice should not be ingested for more than four weeks at a time, however, since it can lead to inflammation of the kidneys. The juice is best combined with other fruit or vegetable juices.

True Watercress can be grown in cool, shallow, slow-moving streams; but it will also grow in very rich and moist soil, albeit not as well. The fastest way to start your Watercress plant is by planting fresh sprigs purchased directly from your local grocer. Select a healthy specimen and place it in a muddy bank near a cool stream. You can also start Watercress from seeds that are planted in boxes in early summer or autumn. When the seeds sprout and attain a height of two to three inches, place them about six inches apart in the mud. You can take cuttings from the plants all summer, but if you do so, don't expect them to flower. And some herbalists say that the medicinal properties are at their peak when the plant is in flower.

FIELD MUSTARD, Charlock
Brassicaceae
Sinapis arvensis

This troublesome weed grows to a height of two feet. Like the other Mustards, it has many-branched stems, but is somewhat hairy.

FIELD MUSTARD
Sinapis arvensis

The leaves are unevenly toothed, and the yellow flowers, which appear in May and June, are followed by the seedpods, which are elongated angular pods that grow vertically along the stem

The seeds of Field Mustard are not as valuable as the other varieties for the making of mustard, but they do contain an oil which is said to be good-burning. The plant is well-liked by cattle and sheep for fodder.

When boiled, the plant is said to make an adequate green vegetable.

GARLIC MUSTARD, Jack-by-the-Hedge
Brassicaceae
Sisymbrium alliaria

As the name implies, this herb gives off a strong Garlic aroma when it is crushed. Garlic Mustard has toothed, heart-shaped leaves and white flowers.

In the past, the leaves of this herb were eaten to induce sweating or to remove obstructions from the digestive tract. Externally, they were said to be good as an antiseptic. The seeds were once used as a kind of snuff and were usually successfully employed to induce sneezing.

The leaves of Garlic Mustard have been an ingredient in certain sauces served with meats, and they have been utilized as salad greens.

GARLIC MUSTARD
Sisymbrium alliaria

It's best to keep cows away from the plant, however, as it tends to lend its disagreeable aroma to the flavor of bovine milk.

HEDGE MUSTARD, Singer's Plant
Brassicaceae
Sisymbrium officinale

Hedge Mustard grows along roadsides and is common in disturbed areas. It tends to accumulate the dust kicked up by passing vehicles, since virtually the entire plant is somewhat hairy in texture. The leaves have deep lobes; they're about six inches long at most. The flowers are tiny, pale yellow, and in clusters; they appear from May through October. The plant grows to a height of about two and one-half feet. The fruit pods are small and hairy, and they grow tight against the stem. They contain yellow seeds that have an extremely bitter taste.

Hedge Mustard was an introduced variety that is now found just about everywhere. The French supposedly named it "the Singer's Plant," since they relied upon it to cure loss of voice. Similarly, an infusion of the herb was said to be a valuable remedy for all diseases of the throat.

HEDGE MUSTARD
Sisymbrium officinale

CACTACEAE
The Cactus Family

The members of this family can be herbs, shrubs, or trees, but the majority are fleshy succulents. Often, these succulents have no true leaves and have prickly hairs. These plants have stems that are flat and cylindrical. The flowers appear one to a plant and can be quite showy. They are usually perfect and regular, but not always. The sepals and petals are often in several series and are not clearly differentiated from each other. Often, they will merge into bracts and are attached at the base of the ovary. The fruit grows into a fleshy berry, which is on the dry side. It is sometimes edible, and it contains many seeds. The family is chiefly of the American continents; it is found from British Columbia in North America, south to Argentina and Chile in the lower hemisphere. The flat shape of the members of Cactacea has evolved into a shape that reduces water loss by cutting down on surface area. The roots are generally shallow, so the succulent stem stores the water.

PRICKLY PEAR
Cactacea
Opuntia humifusa

Brilliant yellow flowers rise from a flat, spiked, succulent cushion of a stem in this species. Prickly Pear often has a reddish center in its flower. The plant spreads out in groups three feet wide and a foot tall. The stems are covered with reddish spikes, so you must approach them carefully. Flowers appear from May to August in sandy, open areas in most of the northern continent.

The species name, *humifusa,* means, in Latin, "spreading on the ground." The untrained eye may see only the dangerous spikes and beautiful and compelling flower of Prickly Pear. But the trained eye can see much more.

Hidden beneath the spiny surface is a multi-faceted gift. The Indians used the thorns as needles to sew their hides. But they could also quench their thirst with the precious liquid found within the stem, and so survive on a desert trek. The juice of Prickly Pear is said to make a cool and acid drink that turns urine red. Strips cut from the stem of the young succulent can make an interesting pickle or be used

PRICKLY PEAR
Opuntia humifusa

as a potherb. The fruit and new joints of the herb can be eaten raw,
boiled, or dried. The seeds are also edible. The fruits of one species,
Opuntia fiscus-indica, were once cultivated to make a thick red syrup
called Quesode Luna.

Indians rubbed the mucilaginous juice on open wounds to dress
them or formed a poultice of the pulp to apply directly to slow-healing
wounds. The stem of Prickly Pear is said to be helpful in rheumatic
pain, ulcers, and gout.

Choose a well-drained, sandy soil in full sun for your Prickly Pear.
Don't try to start it from seeds—it will take too long. A young starter
plant is your best bet. It makes a great indoor plant, but take care not
to overwater.

CAPRIFOLIACEAE
The Honeysuckle Family

Many of the members of this family are blessed with showy flowers. The flowers have a symmetrical construction and form in forked clusters. They form a five-petaled tube with five small sepals. All the parts of the flower are attached at the top of the ovary. The leaves of Caprifoliaceae are often compound; however, they are more likely to be found with simple construction. They are usually opposite. The fruit is quite often a berry but may also be found as a capsule or single seed enclosed in a hard cover (or, in other words, a stone). There are said to be 15 genera of Caprifoliaceae and approximately 400 species. They are found mostly in north temperate regions.

ELDERBERRY, American Elder, Common Elder
Caprifoliaceae
Sambucus canadensis

Renowned for its distinctive flavor as both a jelly and a wine, Elderberry is a common shrub that is fast disappearing from its native

ELDERBERRY
Sambucus canadensis

habitat. It varies in height from three to twelve feet and is unusual in
the Honeysuckle family in that it has pinnately compound leaves and
a distinctive flat-topped flower cluster. Its tiny white flowers are a
delight to the nose in the spring. The stems are covered with a rough
gray bark and, when broken, reveal a soft, woody pulp that can be
removed easily. Elderberry flowers from May to July, followed by dark
purple berries that ripen in the early fall.

The name Elder, roughly translated from the Anglo-Saxon, means
"fire." When the pith is removed from the stems, they can be used as
an instrument to blow air into fires. The genus name, *Sambucus*, refers
to an ancient Greek wind instrument made from the hollowed stem of
the Elderberry. Herbal adventurers are warned, however, that ingesting any part of the herb in excessive quantity may cause symptoms of
poisoning, especially in children. But the cooked berries are a delight
both in pies and the famous Elderberry jam. They were also used as
a black hair-dye and a tincture to color baskets. Native Americans are
said to have used the herb and flowers for their antiseptic properties.
The bruised leaves and flowers could supposedly be used to wash the
skin of diseases and inflammation.

Elderberry bark and flowers were also said to be a purgative, a
diuretic, and a cathartic. The syrup of the berries is said to sooth coughs
and chase away colds. A tea of the Elder flower supposedly has a
calming effect and is a pleasant drink just before retiring. Some Indians
reportedly used a tea made from the root bark to help in childbirth and
congestion. The fresh berry juice mixed with lard or a cream base has
been used as an ointment to heal burns. An infusion of the flowers,
when applied to the face, was said to smooth wrinkles, ease sunburn,
and fade freckles. An Elderberry flower sachet, soaked in the bath, is
said to be a soothing and refreshing way to end the day. The fruit juice
was once thought to be a good remedy for rheumatism. The expressed
juice of the leaves, when rubbed on your clothing or skin, is said to
keep flies away. Sprayed on your garden, the juice is said to keep insect
pests under control. It has been reported that the leaves, when stuffed
in one's rear pants pocket, prevent chafing.

The flower and the berry have myriad culinary uses, each one
more delightful than the next. Elderberry flower wine tastes somewhat
like Muscatel wine. And the flower mixed with lemon rind in boiling
water can be chilled for a delicious summer cordial. The flower tops
can be made into sherbet, dipped in batter and deep-fried, or made
into vinegar, syrup, and the creme-de-la-creme, Elderberry flower jelly.
The Elderberry berry is also made into a scrumptious jelly, tasty chut-
ney, and a thick, spicy sauce.

Elderberry prefers a sunny, moist spot with a fairly rich soil. It can be propagated from seeds but it is much faster to grow Elderberry from a leafless shoot. In the fall of the year remove cuttings and set them in moist soil to form a hedgerow, or along the edges of the garden to attract the attention of passing birds which may otherwise be attracted to your garden. The Elderberry needs a great deal of water to start a successful new growth. Don't try to grow your Elderberry in a small indoor container: the roots need a lot of room to spread.

BLACK ELDER, European Elder, Bore Tree, Elder, German Elder
Caprifoliaceae
Sambucus nigra

Black Elder grows to a greater height than its American cousin. It can form a small tree that approaches thirty feet in height. While Elderberry has yellowish-gray bark and opposite leaves, Black Elder has light brown bark at the bottom of the stem that turns to a gray-white at the top. The bark of Black Elder is rough-textured. The leaves of Black Elder, while also opposite, are, however, a darker green color and broader at the base, with a more saw-toothed edge than those of Elderberry. Black Elder grows in moist, shady places, starting among the underbrush.

In some European countries Black Elder was thought to be closely connected with magic. The "Elder Tree Mother" was supposed to dwell in its limbs and was said to haunt the wood cut from the tree forever. The origin of these superstitions may have been in Biblical times, since the Cross of Calvary was said to have been made from the wood of the Black Elder.

All parts of Black Elder have been used medicinally. The roots and bark, however, are little-used today, since they can cause inflammation of the gastro-intestinal tract and vomiting if taken in large doses. A tea made from the leaves and the young shoots, though, reportedly eliminates excess water from the body and is said to be good for urinary and kidney problems, as well as constipation. The flowers, when boiled in water, allegedly increase perspiration and are used as a cure for colds and rheumatic problems.

Unless you cook the berries, you'll be ill all night after eating them. But they make a great jam or jelly alone, or in combination with other fruits, like currants and crab apples. Elder flowers are said to be a rejuvenator for skin, but be careful if using them, for their expressed juice may stain your clothes.

Black Elder likes moisture and a semi-rich soil. It is best grown from cuttings planted in late fall or over the winter. Cuttings root quickly and can be transplanted.

CARYOPHYLLACEAE
The Pink or Carnation Family

Members of the Pink family are mostly herbs with a distinguishing swollen node that appears on the stems. Flowers appear by themselves or in a branched group. These flowers often have jagged or toothed edges with five sepals, five petals, and five to ten stamens; all these parts are attached to the base of the ovary. The fruit forms a capsule as a general rule. Pinks love the cooler regions and abound there in the form of about 80 genera and over 2,000 species.

MOUSE-EARED CHICKWEED
Caryophyllaceae
Cerastium vulgatum

Fuzzy leaves and hairy stems mark this low-growing plant. As it spreads along the ground, delicate white flowers are formed atop thin stalks. It grows to a height of one foot, but it is often hard to tell this, since the herb tends to lie along the surface of the ground for most of its life. Mouse-Eared Chickweed flowers have five one-quarter-inch

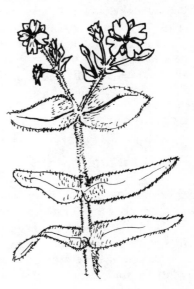

MOUSE-EARED CHICKWEED
Cerastium vulgatum

wide petals distinguished by their deep notches. The leaves form in pairs; they are quite tiny and are often found to have no stalks. The herb is found in waste places everywhere, and the flowers bloom from May to September.

Naturalized from Europe, this herb derives its name from the tiny velvety leaves which resemble the ears of the familiar rodent. Mouse-Eared Chickweed is a sleeper in the garden, where it is often regarded as a pest; but a clever herbalist will take that nuisance, boil it for a few minutes, and turn it into a culinary delight. The tender leaves and stems add zip to salads, also.

Mouse-Eared Chickweed will easily adapt to most soils and is fond of sunny locales, but it will tolerate some shade. By introducing just a few sets to your garden, and keeping them well watered, you should have as much of the leaves as you will ever need. If you don't pluck it too severely, small cylinder-shaped capsules will form on the plant and seed automatically in the fall.

EGYPTIAN SOAPWORT
Caryophyllaceae
Gypsophila struthium

This herb also contains the glycoside Saponin. Its properties are similar to those of Bouncing Bet, but Egyptian Soapwort is generally recognized for its six-inch-long yellowish-white root. It is a perennial that flourishes in most parts of Europe and North America. The plant grows to a height of about two feet.

As with Bouncing Bet, and because of its Saponin content, Egyptian Soapwort has been used in the treatment of syphilis, gonorrhea, and rheumatism; and it was said to be a valuable remedy for various skin afflictions.

BOUNCING BET, Soapwort
Caryophyllaceae
Saponaria officinalis

Bouncing Bet takes its name from the colloquial reference to a nickname for an Old English washerwoman. The plant, in fact, can be crushed and used as soap. The genus name, *Saponaria,* is derived from the Latin *sapo,* meaning "soap." The herb grows to a height of about two and one-half feet along roadsides and waste places. Its stem is relatively smooth and it generally has few branches, which contain

BOUNCING BET
Saponaria officinalis

oval-shaped veined leaves. The flowers, which appear from July to September, are white or slightly pinkish, with five notched petals that seem to droop at the ends. Bouncing Bet is a perennial.

Poisonous saponins contained in this plant can cause severe irritation to the digestive tract. Bouncing Bet is said to stun fish that have eaten it. The herb is said to have no odor and a somewhat bittersweet taste. If chewed, the leaves are said to impart a tingling sensation to the mouth, followed by numbness.

Formerly used in the treatment of syphilis, gout, and rheumatism, Bouncing Bet contains a powerful glycoside that literally dissolves red blood cells. A decoction made from the herb has been used in the past to treat skin irritations, and it was sometimes used to remedy respiratory congestion by acting as an expectorant.

Bouncing Bet tends to form its own little colonies of plants by means of a spreading, underground stem system. It will do well in a garden, though it may be difficult to control.

STARWORT CHICKWEED, Common Chickweed, Starwort, Stitchwort
Caryophyllaceae
Stellaria media

Starwort Chickweed is a sprawling plant with multiple branches

GARDEN CHICKWEED

and a stem distinguished by its single line of hairs on one side. Tiny white terminal clusters of flowers appear with deeply toothed petals, or the flowers may appear by themselves. The sepals of the flowers appear to be longer than the deeply cleft petals. The leaves are opposite, smooth, and oval-shaped. Starwort Chickweed is an early bloomer, first exposing its delightful blossoms in February and continuing to bloom into December. The fruit is a curious capsule that closes around the seeds in wet weather but allows them to escape in the wind when they are ripe.

One unusual feature of this herb is that the leaves tend to fold over the stem as though they were in prayer during the evening hours. Birds like to eat Starwort Chickweed seeds; they also like the young tops of this biennial. The stem is brittle and is easily broken to make a tasty addition to a green salad.

The young leaves of this herb can be boiled and eaten like young Spinach leaves, and they are just as nutritious. Starwort Chickweed is a little on the bland side, though, so it is best to combine it with the leaves of herbs with a stronger flavor.

The whole herb is reportedly useful medicinally. It may be used fresh or dried, but is said to be best collected from the spring to mid summer. Skin irritations are said to be soothed by a poultice made from the fresh Starwort Chickweed plant bruised and applied directly or boiled in water. An ointment of the herb mixed with lard or a petroleum base is said to facilitate a cure of skin diseases or afflictions. A decoction of the herb was often used to regulate the bowels.

Starwort Cickweed is one of the most common plants in North America. You can find it growing everywhere, but if you want to keep a closer eye on it, a few seeds planted in the early spring or some frequently watered sets should provide you with a goodly amount of greens.

STAR CHICKWEED
Caryophyllaceae
Stellaria pubera

Perhaps the most lovely of all the Chickweeds, Star Chickweed, as its name implies, has a white, star-shaped flower that has deeply cleft petals. The flowers are clustered atop an erect stem. They grow from the leaf axils. Two columns of hair line the stems in this species. The five flower petals look like ten because of deep incisions, and the petals are longer than the sepals. The small leaves, about three inches long, appear in pairs that are opposite each other on the stem. Star Chickweed blossoms from March to May.

This more showy member of the Pink family likes rich woodlands or rocky slopes and its uses are similar to those of its cousins.

CHENOPODIACEAE
The Goosefoot Family

Members of the Goosefoot family are distinguished by their succulent stems and leaves and their tiny clusters of greenish flowers. The flowers are noticeable by virtue of their lack of petals. All other flower parts, however, are attached at the base of the ovary. The leaves of these herbs are mostly simple in structure, sometimes toothed or lobed, and usually alternate. The fruit is small and one-seeded. It is said that there are just over 100 genera and about 1,400 species.

LAMB'S-QUARTERS, Pigweed, White Goosefoot
Chenopodiaceae
Chenopodium album

Lamb's-Quarters is a tall, branching plant with stems that are streaked with red. Its diamond-shaped leaves are coarsely toothed, and they attain a length of about four inches. The tiny green flowers grow in spikelike clusters which appear from June to October. The plant grows on cultivated land or along roadsides.

LAMB'S QUARTERS
Chenopodium album

The species name, *album,* which, roughly translated, means "white," is a reference to the silvery color on the underside of the leaves. Lamb's-Quarters is often found growing on manure and compost heaps, and is, therefore, often disregarded as a mere weed. The leaves of the herb are used in the same way as its cousin, Spinach. Lamb's-Quarters is rich in vitamins C and A and was said to be used as a treatment for various skin problems.

The seeds are said to be highly regarded by Indians of the Southwest, who eat them for an energy boost. Also, large quantities of the seeds may be dried and ground into flour to make a kind of bread.

You may find Lamb's-Quarters as an already uninvited guest in your garden; if so, just sit back and enjoy it. To introduce your own plants, however, sow the seeds in full or partial sun early in the spring; and when the seedlings reach a height of about three inches, thin them to a foot or more apart. The leaves are best used when young and tender.

MEXICAN TEA, Ambrosia, Wormseed
Chenopodiaceae
Chenopodium ambrosioides

Mexican Tea is commonly grown today for the aroma of its dried foliage in flower arrangements, but it is also the source of the drug Chenopodium, which is one of the most effective treatments for intestinal roundworms known to medical science. The herb grows to a height of about three feet. Its leaves are alternate, toothed, and oblong in shape. The flowers, which appear from July to September, are yellowish-green and form spikelike clusters atop the plant. The fruit, also, is yellowish-green, bladderlike, and tiny; and it contains the single, shiny black seed from which the active drug is extracted.

The drug, an oil that is extracted from the crushed seeds through distillation, is aromatic, bitter tasting, and somewhat toxic. It should not be used without medical supervision, especially by children, since an overdose can cause severe poisoning or even death. As its name implies, Mexican Tea is indigenous to Mexico, but it has become common in most of the eastern United States. It is grown extensively in Maryland near Baltimore, where much of the oil is distilled for commercial purposes.

In addition to its use as a vermifuge, Mexican Tea has also been used to treat malaria and some nervous disorders. The American Indians were said to have used it as a remedy for painful menstruation.

Mexican Tea, while found commonly along roadsides and in waste places, can also be grown from seeds sown in early spring. The tiny seeds can be scattered in a sunny area with fairly rich soil, and the seedlings that pop up should be thinned to about a one-foot spacing. The harvested plant dries well for flower arrangements.

GOOD KING HENRY, English Mercury, Fat Hen
Chenopodiaceae
Chenopodium bonus henricus

Good King Henry is a tall perennial with leaves in the shape of an arrowhead. The plant sports tiny yellowish-green flowers that have no petals but form in elongated clusters over the stem. Good King Henry would normally stand about a foot and a half high, but it tends to bend gracefully under the weight of its flower clusters. The bright green leaves reach up to four and a half inches in length. The fruit is a small pod with one seed. The herb is found naturally in partially shaded waste areas with good drainage. It is most prevalent along the Atlantic coast to Georgia and from the Great Lakes south and west excluding the southernmost tips of the Gulf states. Also, it is found through the Rocky Mountains northwest to Canada.

In some areas, Good King Henry is still used as a substitute for Spinach, the leaves of both plants being quite similar. The young shoots

GOOD KING HENRY
Chenopodium bonus henricus

of the herb may also be eaten like Asparagus, but they tend to have a slight laxative effect. Part of the species name, *henricus,* is said to be derived from the name of a legendary elf who supposedly had powers of a mischievous nature. The reference was to another plant with poisonous properties that resembled Good King Henry, and was commonly called "Bad Henry." At any rate, the "good" version of the plant was at one time used to feed poultry, thus earning it the alternate common name, "Fat Hen."

Besides the laxative tendencies of the shoots, the herb is also said to be a good remedy for indigestion, and a poultice made from the leaves was once recommended for the treatment of skin sores.

Should you desire to grow this plant, the seeds are best sown in early spring about a foot apart. The soil should be of relatively good quality, and the seedlings should be kept well-fertilized. Leaves and shoots should not be trimmed from the plants until after the first year.

SLENDER GLASSWORT
Chenopodiaceae
Salicornia europaea

This herb is found along coastal marshes and salty waterways and grows to about a foot and a half in height. The leaves are very tiny, scalelike, and opposite. A fleshy succulent, Slender Glasswort has tubular stem joints that are each a little longer than they are wide. The flowers appear from August to November in the hollows of the upper joints. They form a spike of green and appear in groups of three. In

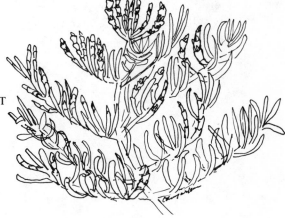

SLENDER GLASSWORT
Salicornia europaea

keeping with its fondness for tidal marshes, Slender Glasswort grows freely along the Atlantic coast from New Brunswick south to Georgia, and is also found in the Great Lakes region. Look for it in salt licks or peat bogs along the Pacific coast as well.

Slender Glasswort signals fall along the marshy shores by transforming itself into a reddish beacon of the changing season.

For those who are tired of the taste of kosher dills, pickled Glasswort is said to be an excellent substitute when steeped in vinegar. Owing to its marshy habitat, Slender Glasswort has a salty taste and can be tossed into a salad for a natural sodium substitute. It can provide an interesting texture to your cheeseburger.

Cultivation of this herb is limited to those areas simulating its natural habitat.

JOINTED GLASSWORT, Mars Samphire
Chenopodiaceae
Salicornia herbacea

Jointed Glasswort is a low-growing herb found in salt marshes and muddy seashores of the continental coast. It is a succulent with smooth, cylindrical branches that are light green in color. No leaves adorn the jointed branches but minute flowers in groups of three form there in the axils.

Jointed Glasswort has a high soda content and was used in the old days to make soap and glass. Hence, the name "Glasswort." Cows are fond of its taste because the stem contains a salty juice. When the plant is young and tender, the stems and branches may be soaked in malted vinegar and eaten as a pickle.

Marshy salt areas are preferable for cultivation.

CRASSULACEAE
The Sedum or Stonecrop Family

The succulent herbs of this family have thick, alternate or opposite leaves. Flowers appear in complex clusters and are usually perfect, meaning that they have both male and female reproductive parts. The sepals wag free or are united in a tube. The petals are free or somewhat united. There are ten or less stamens and may be inserted in the corolla tube. Within the more than 20 genera there are approximately 1,500 species, many of which are used as ornamental houseplants. In some species, plantlets will grow along the leaves, fall to the ground, and root.

HOUSELEEK, Jupiter's Beard, Devil's Beard, Hens & Chickens, Live Forever, Thunder Plant, St. Patrick's Cabbage
Crassulaceae
Sempervivum tectorum

The thick, succulent leaves of this European perennial enable it to withstand lengthy dry spells, since they are adept at retaining moisture. The leaves grow directly from the fibrous root, forming a rosette of sorts ranging from two to four inches in diameter. The purplish leaves are spiny-pointed, fringed with minute hairs, and range in size from one to two inches. The flowers appear on the top of the stem in a cluster of starlike, rose-colored flowers that are located on one side of the

HOUSELEEK
Sempervivum tectorum

stem. They have twelve petals, and they blossom infrequently. The flower blooms on a separate shoot which reaches approximately one foot in height: the shoot has its own leaves.

The genus name, *Sempervivum,* means "always living," and the species, *tectorum,* means "of the roofs or house." In ancient lore, the plant was reputed to be an aphrodisiac. The Romans regarded this herb highly because of its impressive appearance and alleged mystical properties.

Potted Houseleek plants were commonly grown by the doors and on the rooftops of ancient Roman homes, where it was thought to be a cheerful protector against the elements, especially lightning and fire. Inhabitants therein were sure to be prosperous and sheltered from demons.

Herbalists have used Houseleek for a variety of medicinal purposes. It is said to cool the fevered brow and to act as an astringent and a diuretic. The leaves and their juice have a salty, acid taste that will draw the tongue. The bruised leaves of the fresh plant were applied as a poultice to burns, bruises, and skin ulcers. Supposedly, when the juice is mixed with clarified lard and applied to inflamed skin, it will reduce redness and swelling. It is said that, by applying the inner surface of the leaf, warts and corns will disappear from the skin. The juice, mixed with honey, has been used to ease soreness of the mouth from thrush. Crushed leaves were applied to insect bites to ease the sting. The juice was once thought to remove freckles. An infusion of the leaves can be taken internally or a decoction applied externally for shingles, hemorrhoids, worms, and uterine neuralgia.

Houseleek grows well in dry, well-drained rocky soils. Since it grows fast by means of offsets, it is excellent treatment for an old wall or in the rock garden. The flower heads have no odor, and they die soon after the pale red-purple blossoms appear. But do not fear, new offsets will soon take their place. It's easy to propagate Houseleeks by planting offsets at any time of the year. Plant from six to nine inches apart. No fertilizer needed here! It will make a good garden border outside; indoors, treat it like a cactus.Choose a sunny window and soil that is half potting soil and half coarse sand. Keep the soil slightly moist during growth spurts, but allow the plant to rest in the winter by keeping the soil nearly dry.

DIPSACACEAE
The Teasel Family

The herbs of this family are distinguished by their compact flower heads. Their flowers are symmetrical, with five united petals and sepals. The fruits are generally seedlike, with a bristly or spiny exterior. These annual or perennial plants include some 10 or so genera, and about 270 species, most not native to the United States.

TEASEL
Dipsacaceae
Dipsacus sylvestris

Teasel is a tall plant, easy to spot because of its distinctive spiny, egg-shaped head from which its tiny lavender flowers emerge. Long, hairy, lance-shaped leaves form opposite each other on the six-foot barbed stem of this monster. The upper leaves are generally fused together at their bases, surrounding the stem and forming a cup-shaped area that serves to collect rain water. The little flowers begin to bloom in the center of the head around July, and they proceed to spread in

TEASEL
Dipsacus sylvestris

an upward and downward direction on the flower head until early fall, when the head turns brown and fills with numerous dark, rodlike seeds. Teasel is now common in southern and eastern Canada, and most of the northern United States. Old fields are the best places to look for it.

Teasel was once cultivated by weavers, who used the spiny head as a comblike device to tease fresh wool; hence the name. Its seeds are a big favorite of small birds.

Teasel is best cultivated from seeds sown in late spring in a sunny area. It will adapt to just about any soil condition. Allow plenty of room for this tall giant. The flower heads are sometimes used in dried flower arrangements, and if you desire to gather the spiny balls, you should cut them just after they have formed their seedpods, and dry them in an inverted position.

ERICACEAE
The Heath Family

Members of the Heath family are perennials. The flowers are usually symmetrical with four or five petals and twice as many stamens. All of the flower parts attach to the ovary. The leaves are usually thick and tough, sometimes waxy. The fruit is usually a berry or it is berrylike. Botanists say that there are about 50 genera and somewhere around 2,500 species. Plants of this family prefer an acid soil, and they include some common edible berries.

WINTERGREEN, Teaberry
Ericaceae
Gaultheria procumbens

Wintergreen is a creeping, evergreen plant that sends up short shoots from its horizontal stem. The shoots have oval, toothed leaves and they develop white, drooping flowers that spring from the leaf axils. After the flowers come red berries that are edible and have a spicy, or Wintergreen, taste. They are often commonly called Teaber-

WINTERGREEN
Gaultheria procumbens

ries. Wintergreen likes at least partial shade and sandy soil, and it is commonly found wild from eastern Canada south to the Gulf coast.

Wintergreen is best collected late in the fall. Its constituents are commercially extracted with steam. The flavoring has been used for teas, medicines, and, most popularly, chewing gum. Wintergreen is said to be an effective diuretic. The berries themselves are thought by some to be quite tasty.

To cultivate this herb, root divisions of the wild plants can be set in spring or fall, preferably in a well-mulched soil in a shady area.

GERANIACEAE
The Geranium Family

The Geranium family is one identified by its leafy plants with small white or pink flowers. The flowers usually have five petals and stamens in multiples of five up to fifteen. Leaves of the plants in this family can be either opposite or alternate on the stem, and they are both simple and compound in structure. The fruits originate from a long, chamberlike pistil. The seedpods tend to burst open when the seeds are ripe. There are said to be approximately 11 genera and about 800 species in the Geranium family. They are found commonly in a range from southern Canada to the mid-United States.

WILD GERANIUM
Geraniaceae
Geranium maculatum

Similar to Herb Robert in appearance, Wild Geranium has lavender flowers found in clusters of two to five. The flowers have long stalks, are about an inch wide, and have five petals and twice as many stamens. The leaves are about five inches long, have several lobes, and are coarsely toothed. The fruit is a long capsule which splits open when the seeds are ripe and forms five curved strips that remain united at their base. Wild Geranium grows to a height of two feet, and it flowers from April to June. Like Herb Robert, it is found in rich woods or well-drained meadows. Appreciated most for its appearance, Wild Ge-

WILD GERANIUM
Geranium maculatum

ranium can be found in the wild in southern Canada and in the United States east of the Mississippi River.

HERB ROBERT
Geraniaceae
Geranium robertianum

Herb Robert may be recognized best by its disagreeable odor, which is heightened when the plant is crushed. It will grow to a height of two feet, and both the stem and leaves have a reddish tinge to them. The leaves are up to three inches long and have from three to five lobes with coarsely toothed segments. The flowers, which appear from May through October, are usually found in pairs with long stalks. They are pink to lavender in color, about one-half inch wide, and five-petaled. Herb Robert is usually found in rocky wooded areas and moist ravines.

Speculation has it that the plant was named for Robert Goodfellow, otherwise known as Robin Hood, probably because of its supposed beneficial medicinal properties.

Herb Robert was often called "Bloody Mary" by the common folk, because its reddish stems were once thought to be good to staunch the flow of blood from wounds and contusions. A tea made from the herb was said to be good for various intestinal disorders, including diarrhea, gastritis, and enteritis. And a poultice made from the leaves has been used in the past for relief of bladder pains, bruises, and persistent skin problems. The Cherokee Indians supposedly chewed the fiber of the root and then blew it into the mouths of children suffering from thrush (a contagious oral infection).

Herb Robert is an annual plant, best gathered in the wild. Those interested in growing the herb at home would be advised to attempt to simulate the plant's natural habitat, a moist, shady area, for best results. Seeds should be sown in the early fall.

IRIDACEAE
The Iris Family

Brilliant showy blossoms atop long stalks mark the herbs of the Iris family. Under the stem you will often find bulbs, corms, or rhizomes. All of the flower parts are attached to the top of the ovary, and consist of three petals, sepals, and stamens; flowers are usually symmetrical. The leaves overlap each other at the base, and are somewhat parallel. The shape of the leaf is simple, but leaves are folded and alternate on the stem. There are an estimated 60 genera and about 1,500 species, growing mostly in temperate and tropical regions.

SAFFRON, Crocus
Iridaceae
Crocus sativus

Saffron, the purple herald of winter, is one of the latest bloomers in the garden. At one time it was used extensively for its yellow dyeing and flavoring properties. The huge volume of stigmas and styles from the Lily-shaped blossoms needed to produce a small amount of the oil has made it impractical for use by many, and expensive for the rest. Saffron is a small plant, growing to a height of only four inches. One precious blossom forms on each stem. The leaves are very slender and stay green through much of the winter in many areas.

Saffron has a bulbous root called a corm. Unlike some other herbs that open and close at various intervals during the day, Saffron blossoms stay open all night, and can be enjoyed on a midnight stroll as much as on a midday jaunt. Saffron has been cultivated from ancient times.

Saffron tea steeped in brandy was said to chase the measles away. It also causes sweating beneficial to fever reduction, and, according to some old herbalists, it was effective in removing the flatulence associated with children's colic.

Pant the corms early in the spring in light, but well-drained soil. Choose a spot that gets plenty of sunlight and set the corms in rows at least two inches apart. Cover the corms with about three inches of soil, and water well at first. The plants will reproduce in this manner most effectively. You might try experimenting with mechanical pollination to create a viable seed.

SAFFRON
Crocus sativus

IRIS
Iris versicolor

IRIS, Blue Flag
Iridaceae
Iris versicolor

Striking blue flowers with interesting veins are the most noticeable thing about the majestic Iris. Rising to a height of three feet, the erect stalk supports the Orchidlike flowers. The enchanting blue of the flowers is broken up occasionally by streaks of white and by the yellow of the sepals. The flowers grow to four inches in width. There are three non-bearded sepals, three erect, upright petals, and two arching styles along with one other style. The leaves are thin and swordlike and are a pale green color. When the flowers wither, a three-lobed capsule appears: it is the fruit.

Insects are attracted to this highly prized garden plant. The rhizome, which is the official medicinal portion of the plant, grows in a horizontal fashion. It has many joints; it is brown on the outside, but gray within. The odor is faint, but can be nauseating to some, and it has a biting taste. A powder is listed in pharmaceutical dispensatories as having purging powers.

A closely related species, *Iris germanica florentina,* known as

Orris, is cultivated for its fragrant roots. The rhizomes are powdered and used as a fixative in potpourris and many perfumes. It is valued for its violetlike scent. Orris root was a favored source of perfume for ancient Greeks and Romans and gained notable stature sometime later in Europe, where it became known as the fleur-de-lis of France. The root of Orris is harvested by peeling it and leaving it in the sun to dry. The fragrance of the knotty rhizomes develops over a long period of time, so they must be stored for at least two years. An old dentifrice had Orris root as its base. It was made of 1 ounce of powdered Orris root, ½ ounce of Myrrh, ½ ounce of powdered Kino, and 2 ounces of Cream of Tartar, mixed well.

Both of the above herbs may be cultivated by planting in ordinary well-drained soil; they'll do even better in rich soils. The rootstocks should be set six inches apart in the spring, under two to four inches of soil. Many nurseries carry them; don't expect flowers the first year, but they should appear in the second. The plants need lots of sun; they'll provide you with beauty and fragrance.

LABIATAE, or LAMIACEAE
The Mint Family

Many of the members of this family are quite well known for their refreshing aroma and taste. Herbs in the Mint family are usually annuals or perennials, and they are found all over the world. Some Mints are grown as garden ornamentals. Their stems are normally square-shaped, and the leaves can be simple, opposite, or whorled. The flowers have five petals which fuse into a two-lipped tube. The fruit is a four-lobed nutlet, each lobe containing one seed. There are said to be around 180 genera and as many as 4,000 species.

WOOD BETONY, Lousewort
Labiatae
Betonica officinalis

Wood Betony grows to a height of three feet and is distinguished by its heart-shaped, downy leaves. The leaves are toothed and deeply wrinkled. The stem is also hairy, four-sided, and rarely produces branches. A spike of dense purple flowers forms from June to August.

WOOD BETONY
Betonica officinalis

The seeds are three-sided and sheltered in brown pods that fall from this aromatic plant in the fall.

The common name, Betony, at one time meant "surprise." Wood Betony is said to be endowed with the power to keep away evil spirits and bad dreams. You might try to stuff a small pillow with the herb and toss it into your bed to see if it works. The leaves can be used as snuff and as an additive to smoking mixtures.

The tuberous roots of Wood Betony are gathered in the fall of the year for their supposed food value. It is said that you can eat them just as they are or roast them for a tasty appetizer. If it's the leaves you're after, soak them in lightly salted water for a few minutes before eating. You might try a few handfuls steamed with sautéed mushrooms and onions.

At one time, Wood Betony was thought to be useful against venereal disease, but its effectiveness has largely been disproved. The ancient Aztecs used the herb to take away fear, as did European herbalists. The whole herb can be used for a variety of cures. It is recommended that the herb be collected in midsummer, and then dried before use. Wood Betony was once used for cures of the head. It was thought to be a tonic and nervine. Headaches are allegedly wiped away, along with watery eyes and earaches. Add a little of the leaf to wine, and it is said to make a gargle that will dull the pain of a toothache or heal the bite of a mad dog.

Wood Betony grows in most places from southern Canada to the United States, except for the Gulf Coast, which does not have dry enough winters. Select a fairly rich soil that is well-drained and in a reasonably sunny location. This perennial takes two years to mature, so starting from seed can be a slow process. You might try root divisions in the fall or early spring, and then trim back the flwer spike after the blossoms die to prolong the flowering season. Divide the root clumps every several years so that they don't choke out the colony.

DITTANY, Maryland Cunila, Stone Mint, Wild Basil
Labiatae
Cunila origanoides

Dittany is a bushy plant that grows in soil that tends to be a little acid. A clean, sweet smell permeates the entire plant, and bees are attracted to its pinkish-purple flowers. The herb grows naturally in woodlands and thickets. This native American species was used to purge the body of the common cold. Dittany was used by early Eu-

DITTANY
Cunila origanoides

ropean settlers in the United States for its diaphoretic properties and its pleasant taste as well. When brewing, Dittany is said to resemble sweet Thyme and Bay in smell.

AMERICAN PENNYROYAL
Labiatae
Hedeoma pulegioides

This American native was highly regarded by the Indians for a variety of medicinal and culinary purposes. The plant curves slowly upward from a horizontal position, displaying fine silver-gray hairs along its square stem and oval-shaped leaves. Often, on a lucky walk through the woods, one may come upon a sweetly scented hollow where a carpet of American Pennyroyal grows to a foot in height. Tiny bluish flowers bloom where the leaf joins the stem. The entire plant has a refreshing aroma and can be brewed into a soothing and full-bodied tea.

American Indians introduced the herb to early settlers, and it is said they used it as an aid to reducing fevers. If the tea didn't break the fever, a poultice of mashed Pennyroyal and salt was smeared over the body. The entire family benefited from this herb: to rid baby of colic, a few drops of the tea are administered; women drank the hot

AMERICAN PENNYROYAL
Hedeoma pulegioides

tea to ease menstrual pains; and strong doses of the tea were prescribed to rid one of head colds. A few drops of the oil of American Pennyroyal was said to make a good insect repellent.

American Pennyroyal will reseed itself, but a faster way to begin your tea plants is by setting root cuttings in a shady area with plenty of water. If taking the plant from the wild, be sure to cut off the leaves in the fall so all the strength is concentrated in the roots. The roots should soon put out new sprouts.

HYSSOP
Labiatae
Hyssopus officinalis

Fragrant purple flower spikes of this foot-and-a-half tall herb are noticeable for their pleasant aroma. Hyssop blooms over a long period of time. The square stem rises from a woody base. The plant blooms from June to October. The leaves are dark and flat, but are sharp-tasting with a sweet smell like Sage.

The Bible mentions Hyssop frequently. In the Book of John, Hyssop was said to be used along with vinegar on a sponge to quench Jesus' thirst as he hung on the cross. Priests were said to have used it to cleanse lepers and it was dipped in the blood of the lamb on

HYSSOP
Hyssopus officinalis

Passover. The perennial herb found a home in the state of Michigan, where its essential oils were used as a base for perfumes. Trappist Monks used Hyssop to flavor their Chartreuse liqueur. Its minty flavor and rich aroma led it to be used as a meat preservative.

Hyssop will ease the pain of a sore throat, and it will soothe sores and heal snakebites. It was used as a cure for rheumatism, and a hot tea loosens the phlegm in the throat of those with a cold. It was used in combination with Marigold to rid children of the measles. The leaves in a poultice, applied to bruises, remove pain and discoloration, supposedly. A warm tea is said to be useful against hoarseness, insomnia, jaundice, and hysteria. Hyssop tea will increase the circulation while lowering blood pressure, according to some herbalists. It is said to take the sting out of bug bites, and it has been said to kill lice. To remove discoloration from one's eyes, Hyssop soaked in a cloth can be applied directly while you lie down for about half an hour.

Hyssop leaves make a bitter condiment and are sometimes used in soups or salads.

Hyssop can be grown from seeds in the early spring. Choose a well-drained soil in partial shade. The plant grows best where there is a cool winter, and it should be thinned to about a foot apart. It is a slow grower, so you can get faster results by using commercial seedlings. Stem cuttings in the spring can be reset to produce new plants. Hyssop conforms well to a window box, but it also makes a nifty hedge.

After three to four years, Hyssop should be replanted either in the spring or the fall because the herb tends to become woody after that.

ENGLISH LAVENDER, True Lavender
Labiatae
Lavandula officinalis, L. vera

Today, Lavender means two things to most people: a soft purple color and a sweet, assuaging scent. English Lavender, or True Lavender, grows to a height of three feet. It is a gray-leaved plant with small, opposite, linear, blunt leaves that are hairy. The narrow leaves grow from a crooked, multi-branched stem flaked with a gray bark. The purple flowers are tiny and appear in rings; they form a terminal spike. A highly prized oil is found in the calyx of the flower. English Lavender is not a showy plant, but it holds a multitude of splendors in its modest form. It flowers from July to September. The seeds form in the terminal spikes after the leaves have faded.

English Lavender gets its name from the Latin *lavare,* meaning "to wash." Its name refers to the scented baths and soaps the Romans relished before the days of Christ. Lavender is said to symbolize distrust, based on an ancient belief that the asp that killed Cleopatra lurked underneath the herb's dusky branches. According to Biblical legend, Mary hung the baby Jesus' clothing on the gray Lavender bush, where

ENGLISH LAVENDER
Lavandula officinalis

the sun and the wind gently dried them. When Mary returned for the dried articles, the bushes were laden with the sweet scent we know so well today. The poet Dion Clayton Caltrop says of Lavender:

>   Here lies
> Imprisoned in this gray bush
>   the scent of
> Lavender.
>
>
>   It is renowned for a simple purity
> A sweet fragrance and a subtle
>   strength it is the odor of
> the domestic virtues and the
>   symbolic perfume of a quiet life.
>
>
>   Rain
> Shall weep over this bush
>   Sun
> Shall give it warm kisses
>   Wind
> Shall stir the tall spikes
>   Until such a time as is required
> When it shall flower and so
>   Yield to us its secret.

English Lavender is used as an insect repellent, especially for mosquitoes, midges, and flies. It is also said to kill lice and repel moths. The herb was highly regarded as an aphrodisiac and a stimulant. Few could soak their feet in a hot bath of Lavender without losing all feelings of fatigue and melancholy. Of all the varieties, English Lavender is probably the most widely known because it has the highest quality of fragrance. It is the oil of the flowers that is used medicinally, as well as in a culinary and cosmetic vein.

English Lavender oil has been applied to tumors, sores, and snakebites, as well as sprains and varicose veins. It is said to ease the pain of rheumatism and calm the nerves so as to prevent nausea and neuralgia. A gargle of distilled water of Lavender assuages hoarseness, hemmorhaging, and hysteria. Supposedly, some cases of mental depression are relieved when the oil of Lavender is rubbed into the temples. Old herbalists credited it with the ability to lift a melancholy spirit, restore sight, and increase the memory. The flowers and leaves of English Lavender are combined to make a facial steam to cleanse the skin. The herb is a big item in potpourris, sachets, and pillows.

English Lavender is a hardy grower over most of the continent, but it needs protection from winter's chill. New plants may be begun by cuttings, root divisions, and from the slow-sprouting seeds. Stem cuttings can be taken at any time, but root cuttings do best in the spring. The first-year flowers of this plant should be trimmed back in order to make a bushier plant. Established plants may be left to grow but they tend to get woody, and this part of them should be removed. The sets like well-drained, limed soil. Choose a sunny location guarded from the wind. Spring is the time to trim back your plants from year to year, and there is no better plant to make knot gardens out of. To harvest the herb, cut the stem six inches from the soil just as the buds begin to open. Hang it upside down until completely dry in a warm, dry place. Remove the crumbly leaves and seal their freshness in an airtight container. English Lavender makes a suitable indoor plant that will add a soothing aroma to any room as the North Wind howls round the chimney.

SPIKE LAVENDER
Labiatae
Lavandula spica

Spike Lavender differs from English Lavender in that the leaves are broader and shaped more like a spatula. The flower cluster of Spike Lavender is more compressed and the bracts from which the flowers grow are much narrower. The herb yields several times more essential oil than English Lavender, but it is not as good a quality. Some say the odor of Spike Lavender smells more like Rosemary, and thus is called "Lesser Lavender." Great care is taken not to gather Spike Lavender when harvesting for oil distillation. The herb prefers a sandy soil with some shade, in an area that is protected from the wind. Again, use the leaf cuttings or root cuttings for quick propagation; and grow from seeds if you have a month and a half to spare.

FRENCH LAVENDER
Labiatae
Lavandula stoechas

This herb forms tiny, dark violet blossoms featuring bright colored leaflets that protrude from the terminal end of the flower cluster. The

FRENCH LAVENDER
Lavandula stoechas

leaves are narrow and grow in whorls along the length of the stem. Unlike its cousin, English Lavender, French Lavender never grows in limestone soil: it prefers the sandy coastal regions. The name comes from the Romans, who named an island in honor of the plant. French Lavender was used as a bath and for strewing on church floors on festive occasions. It was also used in bonfires to ward off the Evil Ones. The herb was said to be one of the ingredients in Four Thieves' Vinegar, a concoction said to protect scavengers of the dead from contracting the plague.

PEPPERMINT
Labiatae
Mentha piperita

Peppermint is a common garden escape in America and many European countries. It is well known for its distinctive flavor; it is the source of the all-important volatile oil. Peppermint grows to a maximum height of four feet under ideal conditions. It is a hardy perennial, and it has reddish stems that are typically square. The leaves are small and finely toothed. The flowers, which bloom in late summer, appear in dense spikes at the top of the stem. They form a sort of whorled cluster

PEPPERMINT
Mentha piperita

of light purple blossoms. The foliage of Peppermint is distinctively aromatic, and it has a first pungent, then cooling taste if chewed. The cooling effect comes from the menthol contained in the volatile oil of Peppermint.

Historians tell of the use of Peppermint by the Greeks and Romans, both for cooking and for a decorative touch at ceremonial events. Reportedly, its medicinal uses were not fully discovered until mid-18th century Europeans did some experimenting. Because of its current popularity as a flavoring and medicinal ingredient, Peppermint is widely cultivated in England, France, and the United States.

Peppermint oil is said to have a valuable antispasmodic action when ingested. The herb has, therefore, gained quite a reputation for its ability to soothe the digestive tract, whether the problem might be simple indigestion, flatulence, colic, diarrhea, or even cholera. It is said to anesthetize the nerve endings of the stomach, thereby making it an effective preventative of seasickness. Also, the menthol found in oil of Peppermint is said to be an effective analgesic used to treat minor pain associated with certain neurological disorders. Some herbalists warn, however, that the continual use of this herb may be harmful.

Besides its most common use as a flavor for chewing gums, Peppermint has been employed to flavor candies and other confectionery items. The herb has long been used as a common household tea, said to be invaluable as a remedy for the annoying symptoms of the common cold.

Peppermint can be found growing wild in most moist areas, including ditches, marshes, and along streams. It is commonly noticed growing around homes, especially along damp, shaded basement walls. The herb is ridiculously easy to cultivate. Any moist, well-drained soil will do. Root cuttings set in the early spring should be quite successful. Thin the young seedlings to a foot or two apart, and be sure to keep the weeds away completely, since almost any competitive plant will detract from the precious flavor of your herb. The plant should be harvested in late summer, just before flowering. Choose a dry, sunny day for cutting, and if you wish to extract the oil, the distillation process should be carried out immediately. Otherwise, the herb may be dried in the usual manner, in a cool, dry place out of the sun.

ENGLISH PENNYROYAL
Labiatae
Mentha pulegium

This native of Europe and Asia is known for its ability to repel mosquitoes and, particularly, fleas. The species name, *pulegium*, means "flea." English Pennyroyal is a creeping, flat herb. The flower stalk, bearing purple blossoms, is the only visible structure of the plant that rises from the surface of the earth. It is a hardy perennial that can grow up to six inches in height. It has a strong flavor, and it remains green throughout the year. The leaves are about one-half inch long, opposite, and oval-shaped. The herb blooms from mid to late summer.

English Pennyroyal was grown on ancient sailing ships to clean casks of drinking water that had stagnated with age. Pennyroyal tea was often enjoyed for its flavor, but is also known for its ability to remedy colds and menstrual problems.

At one time, Pennyroyal was used in the kitchen for more than its flavorful tea. The leaves have a strong flavor and can be used for soups, salads, and stuffing; it should be finely cut, though, because of its strong flavor. It fell out of favor for puddings and sauces because of its slightly bitter taste.

English Pennyroyal should not be used by pregnant women, as it causes uterine contractions. It is said to relieve hysteria, nausea, and other nervous problems. The herb is said to be good for colic and flatulence in children. It is reputed to break up congestion, and when applied as a poultice, it was said to draw the bruises from your skin.

In a moist, loamy soil, the seeds will sprout in the spring. You can also propagate the plant by dividing old rootstocks in the spring. Allow six inches between the plants and keep them well-watered. Pennyroyal makes a pleasant house plant, and it might keep the mosquitoes from your door. The herb makes a fragrant carpet when allowed to grow along a walkway. You can pick the leaves at any time, but they're best just before the flowers open.

SPEARMINT
Labiatae
Mentha spicata

Spearmint is said to have a much stronger flavor, but much weaker medicinal properties, than Peppermint. It, too, is a very common garden escape that may have been carried to the United States by the Pilgrims. Spearmint has widely creeping root runners, and it is slightly shorter in stature than Peppermint. Its square stems will usually grow to only two feet in height. The herb's two-inch-long leaves are smooth and deeply veined, and, like its cousin Peppermint, Spearmint develops

SPEARMINT
Mentha spicata

pinkish or purple whorled flower spikes at the top of its stem.

Ancient references to the payment of tithes in the form of Spearmint and certain spices suggest that this herb was well regarded for centuries. Old herbalists wrote of its beneficial effect on the digestive tract as well as its delightful scent as a perfume. Curiously, mice and rats are said to loathe its aroma, and the presence of Spearmint is therefore suggested to prevent infestations of the pesky rodents.

A tea of Spearmint is held to be quite beneficial for children's complaints of colic or nausea, but the herb is more preferred for its culinary applications.

Spearmint is used abundantly as a flavoring for chewing gums, toothpastes, and candies. It is also a popular ingredient in jellies, mint sauces, and vegetable and meat dishes. The ancient Greeks were said to have thought it to create a craving for meat. Spearmint also goes well with all fruits, fruit juices, and fruit salads. It is the essential element in a mint julep.

Spearmint grows best from root division set in early spring. They should be placed about a foot apart in moist, shaded soil for best results. Manure is not recommended to enrich the soil, since it tends to increase the susceptibility of Spearmint to a fungus-caused rust which is hopelessly incurable once established. Spearmint should be harvested just before flowering, and on a dry day. The foliage should be dried immediately in the usual manner.

BEE BALM, Bergamot
Labiatae
Monarda didyma

The moplike crimson flower of Bee Balm is its most distinguishable feature. It was discovered to make a pleasant tea for the Colonialists after the Boston Tea Party. A perennial that grows to a height of three feet, Bee Balm is best used as a border plant in your herb garden. The square stem culminates in a thick but rounded cluster of tubular flowers. Reddish bracts form just under the flower cluster. The toothed dark green leaves grow up to six inches long. They are lance-shaped and form opposite each other on the stem. The boon of honeybees and hummingbirds, Bee Balm grows in moist woods and along streams; it flowers from June to August. The herb is found throughout the United States and on the east and west coasts of Canada.

The fragrant florets are a garden escape said to have been named for the Spanish physician and herbalist, Nicholas Monardes, who wrote of them in the 16th century. Bee Balm makes a great addition to bouquets. The name Oswego Tea comes from a New York Indian tribe which used the leaves to make the delicious and soothing tea. Early settlers followed suit, and Oswego Tea became a favorite among rebellious patriots during the boycott of British tea. The entire plant has a citruslike aroma.

BEE BALM
Monarda didyma

The Indians drank Oswego Tea to stimulate heart action, quell chills and fevers, and soothe sore throats. They inhaled vapor from the boiling leaves to clear up bronchial problems. The boiled leaves were then used as a forerunner of Clearasil℠ to clear skin blemishes. The oil from the herb was said to make one sweat. It was also used as a base for perfumes.

In addition to its medical uses, the American Shakers added the leaves of Bee Balm to fruit cups, salads, and jellies.

To harvest the leaves, gather the plant before or after it flowers. Run your hand down the length of the stem as it hangs in an inverted position, pulling the leaves with you as you go. Place them on a screen in a warm, dry place until they are absolutely dry, then quickly bottle them in an airtight container. Bee Balm likes a whole winter growth season and lots of sun in a rich moist soil. Sowing from seed takes a long time: almost two years to flower. So, you can either buy seedlings or take root cuttings from the previous year's growth in the early spring. The fibrous roots are easily pulled apart; all the energy is in the outer portion, so that is what you should replant, about a foot apart. Bee Balm is a shallow rooting plant. For larger blooms, cut back the plant before it blossoms the first year. Following that, removing early blooms for a bouquet will stimulate more blossoms.

WILD BERGAMOT
Labiatae
Monarda fistulosa

This member of the Mint family has a cluster of tubular flowers at the top of a square stem that reaches a height of four feet. The flowers appear from midsummer to early fall in dry fields. Wild Bergamot prefers limestone soils. The pinkish-blue flowers are small, about an inch long; each one has two lobes on the upper lip and three wider lobes on the bottom lip. The leaves are long, deeply toothed, and a gray-green in color. It thrives in most places in North America.

The leaves are crushed and steeped in boiling water to make a soothing tea. It was said that, by breathing the vapor of water and Wild Bergamot leaves, respiratory problems would be alleviated. This tall plant would probably best be placed in the rear of your herb garden.

Wild Bergamot is cultivated from seed or root cuttings. It likes a moist soil and plenty of sun.

WILD BERGAMOT
Monarda fistulosa

CATNIP
Nepeta cataria

CATNIP
Labiatae
Nepeta cataria

Sure to drive cats into ecstasy, Catnip also makes a soothing night-time tea for us two-legged creatures. One of the more well-known members of the Mint family, Catnip grows to a height of about three feet. Pale white or lavender flowers appear from June to September. They form terminal clusters on the main stem and branches. The flowers are tubular, about a half inch long, and have a hairy calyx. The stem is square; the leaves grow to about two and one-half inches long, and they are opposite, triangular, and coarsely toothed. A native of Asia, the plant now grows well along roadsides, pastures, and in waste places. It is found throughout North America.

Catnip contains a chemical that repels insects, known as nepeta lactone. It is this same chemical that may cause the delirious cavorting of our feline friends. In the summer, bees gather around the fragrant flower spikes. And in the winter, finches stop to dine on the brown seeds.

Catnip makes a soothing tea for infants that is supposed to stop colic. A fomentation or poultice of the leaves is good to reduce swelling, according to some herbalists. The tea of Catnip is also said to lull both children and adults into peaceful sleep. It has been used to cure laryngitis, hypertension, and hysteria. It makes one perspire freely, and those suffering from itchy scalp have been known to benefit from its use as a hair rinse. When making an infusion of this herb, herbalists advise that you not boil the leaves, but merely keep a tight lid on them while they are steeping.

Catnip prefers a rich, moist soil in partial shade, but it can handle full sun if necessary. Sow seeds in the fall where you want the plants to grow, but thin them to at least a foot apart. Another way to start the plants is through root divisions in the spring of the year. Catnip likes a lot of water through the dry season. Pinch back the first flower buds for a bushier plant. You can also grow it in the house as an indoor plant if kitty will stay away from it.

BASIL, Sweet Basil
Labiatae
Ocimum basilicum

No kitchen would be complete without Basil. The bushy plant grows to a height of about two feet. The leaves are dark green and

BASIL
Ocimum basilicum

very aromatic. They're slightly serrated and glossy; the sides form a V-shape that serves to collect rainwater. Oil glands on the surface of the leaves create the shiny appearance, while the leaves are somewhat downy underneath. The leaves are in pairs and opposite, and they're placed at right angles to the pair above them. The flowers of Basil appear in racemes in July and August. They are fairly inconspicuous and of typical Labiatae structure. The flowers form a whorl on the leafy raceme; only a few of them are open at one time.

Basil is popular in Italian dishes; its pungent flavor begs to marry with tomatoes and pasta. Europeans have long taken the herb to be a symbol of love. But it has a practical side, too. Basil is said to be good to repel flies and mosquitoes. On the other hand, Basil is also taken to be a symbol of hatred by some European herbalists, who said it had the eyes of a basilisk, which was a legendary reptile renowned for its fatal glance and breath. The common name, Basil, means "the base," or "bottom." The herb is often associated with death. In Persia, it's grown over burial grounds, and in India, where it is sacred, the Basil leaf is placed on the breast of the dead before burial. The oil of Basil is used for perfume. In the southwestern United States, Basil is carried in the pocket in the hope of attracting money. The Indians used it in various ways, including the treatment of snakebite and consumption.

Basil is said to be a stimulant, and it is used to remedy headaches and some nervous disorders. An upset stomach was thought to be calmed by a Basil tea.

Basil is a versatile culinary additive. An aromatic vinegar can be made by soaking Basil in a covered jar for at least three weeks. But you can't go wrong by adding Basil to soups, salads, butter, cottage cheese, vegetables like peas, potatoes, and asparagus, and numerous sauces and salads.

In the garden, Basil makes an attractive border plant. The soil must be rich and well-aerated. Be sure to wait until the danger of frost has passed, otherwise, sprout seedlings indoors under one-quarter of an inch of soil. Seedlings may be thinned to about a foot apart when they reach a height of three inches. Basil will transplant easily after the temperature reaches an average 70 degrees. For bushier plants, rich with leaves, pinch back the tops when they reach a height of six inches. The leaves may be harvested at any time and a new plant will grow if severed at the stem more than six inches above the ground. Basil leaves can be kept fresh for the winter months by quickly freezing them or by storing them in olive oil. Don't be alarmed if the leaves turn black in the oil. If you prefer to dry the herb, sever the stem just as the plant comes into flower. Hang it upside down in a dry, warm place out of the sun. When the leaves are thoroughly dried, strip them from the stem and place them in airtight jars. If you want to grow the plant indoors, it will need at least five hours of direct sunlight or twelve hours of artificial light to do well.

SWEET MARJORAM, Knotted Marjoram
Labiatae
Origanum marjorana

This small perennial has much in common with its relative, Wild Marjoram. It grows to about ten inches high, and its multi-branched stem contains rounded, paired leaves and small clusters of white flowers. The blossoms, which appear in July, are so tightly formed as to resemble knots; hence, the alternate common name.

Some cooks are said to prefer Sweet Marjoram to the Wild because it has a slightly milder flavor. Otherwise, it is quite similar. Not long ago, a salve was made from the herb that was said to be an effective remedy for congestion, especially in children. It was also thought to have some use in the treatment of digestive, respiratory, and neurological complaints.

Sweet Marjoram can be propagated by seeds or stem cuttings. Seeds are best sown in early spring in light, warm, well-drained soil. The seeds are somewhat slow to germinate, and young seedlings must

SWEET MARJORAM
Origanum marjorana

be kept out of competition with accompanying weeds. The plants need about eight inches of clear space around them. Sweet Marjoram should be harvested just before blossoming in July, and the leaves should be dried in a cool, dry, dark place.

WILD MARJORAM, Oregano
Labiatae
Origanum vulgare

Wild Marjoram could also be called "true" Oregano, since it is, in fact, the flavoring originally used in pizzas and other Italian dishes. Today, however, the common grocery store Oregano sold commercially is actually *Origanum mexicana,* a Mexican species of Wild Marjoram. But we know the difference, and now, so do you. Wild Marjoram is a creeping perennial with multiple stems that grow to about a foot in height. The stems contain paired leaves and clusters of pinkish-purple flowers which appear from June through August.

The genus name, *Origanum,* roughly translated, means "mountain joy," a suspected allusion to the herb's striking appearance in its native habitat. Wild Marjoram's medicinal reputation dates back to the ancient Greeks. They reportedly used it as an antidote for poisons and as a remedy for certain neurological disorders. The herb was said to protect

WILD MARJORAM
Origanum vulgare

against the spells of German witches and to drive away ghosts and goblins.

Wild Marjoram is said to be a mild stimulant, and the volatile oil extracted from the herb is touted as an effective painkiller for toothache. It was once thought to speed the onslaught of measles if given in the form of a mild infusion; the same infusion was said to be an effective treatment of intestinal distress.

The herb is a flavorful addition to just about any food, including meats, eggs, and cheeses. Wild Marjoram can be added to butter as a garnish for vegetables, or to cream cheese to make a sandwich spread.

Wild Marjoram can be grown from seeds in a sunny area with light, well-drained soil. The plants can be started indoors until the advent of warmer weather, and they can be kept there, for that matter. Outside, they need about ten inches of breathing space. The herb can be started with root cuttings, also. Leaves should be harvested just before the plant flowers, and drying, which improves the flavor of the leaves, is done in the usual way.

WOUNDWORT, Heal-All
Labiatae
Prunella vulgaris

This common perennial is found throughout the United States and

WOUNDWORT
Prunella vulgaris

southern Canada. It has a creeping rootstock from which arises a square, hairy stem. The leaves are slightly toothed, opposite, and oval-shaped. The fruit is an oval, smooth nut that follows a dense spike of purple flowers that appears from May to October.

A teaspoonful of Woundwort in a pint of brandy, steeped for a few days, does wonders for a sore throat, reportedly. The whole herb, when chopped and soaked in cold water, is said to make a refreshing but somewhat bitter tea. The tea can soothe internal wounds and is said to be good for the treatment of thrush. Woundwort has been employed to stop fits and convulsions. Externally, a wash will ease the pain of wounds and is said to be a good astringent.

Woundwort grows well in most soils, in partial sun. It will reproduce itself from its underground stems or by seeds.

ROSEMARY
Labiatae
Rosmarinus officinalis

This aromatic herb has a colorful history and a multitude of uses. It has long been known as the "herb of remembrance." The ancient Greeks were sure that the plant would improve the mind and strengthen the memory. Rosemary has thus become the symbol of fidelity for lovers, and it is displayed ceremonially at weddings, festivals, and funerals. Rosemary has a sweet smell, and it makes a valuable addition to any herb garden. It can grow to a height of six feet under ideal conditions and thus makes an excellent border plant. Rosemary is somewhat shrubby, with short evergreenlike leaves that are dark green in color and shiny-surfaced. The small blue blossoms, which appear beginning in April or May, form short racemes at the top of the stem. The leaves are more fragrant than the flowers; the aroma is somewhat camphorous.

Rosemary's flowers are said to have taken their color from the cloak of the Virgin Mary when she placed it over a Rosemary bush to dry. The herb has been burned in hospitals in the belief that it would disinfect the air. A concoction called Hungary water, said to have been used with success by a 13th century queen of that country, was thought to be restorative of paralyzed limbs. It was simply a distilled liquid produced from a combination of Rosemary flowers and wine.

Medicinally, Rosemary is thought to be a great stimulant and a remedy for many neurological disorders. An infusion, or Rosemary tea, is said to relieve cold symptoms and headaches. Eating the leaves

ROSEMARY
Rosmarinus officinalis

is said to be a sure way to restore lost appetite; and ground into a powder, the leaves, when rubbed over the body, are said to make one feel light and merry. Rosemary is thought to be effective in stimulating the growth of hair and thus preventing baldness. It is also said to make a good shampoo for dandruff.

The herb is a tasty additive for lamb dishes, but it goes well with many meats, even fish. Like the other Mints, Rosemary can be added to jellies, fruit salads, and wines. It can also lend an unusual flavor to eggs, cheeses, and cookies.

Rosemary grows very slowly from seed, so it is most often started from stem cuttings. The herb prefers a dry soil situation in bright sunlight. It should be sheltered while growing, perhaps by a low wall with a southern exposure. The cuttings should be taken in late summer, and they can be started in sandy soil. The plants should be spaced about three feet apart. Leaves may be harvested anytime, but they're best just before the flowers bloom. They can be dried in the usual manner in a cool, dry area out of the sun.

BLUE SALVIA
Labiatae
Salvia azurea

This species of the *Salvia* genus is descriptively similar to the

BLUE SALVIA
Salvia azurea

others. It has blue flowers appearing in a spikelike cluster, and it has the typical square stem and textured, downy leaves. It grows to a slightly greater height, about five feet, and usually flowers from July to October. A perennial, it is commonly found in the southeast to midwest United States, usually in open pastures or dry prairieland.

Blue Salvia will make a colorful addition to your herb garden; it is best suited for a border position.

SALVIA, Scarlet Sage
Labiatae
Salvia coccinea

Another decorative species, Salvia is recognized by, you guessed it, its spectacular red flowers, which form on an elongated, rather sparse-looking spike. The large flowers are typical of the genus, as are the leaves of the plant. Flowers appear from late spring to early fall. Salvia prefers a sandy soil; therefore, it is usually found in coastal areas from South Carolina south to the Gulf coast states.

Salvia mellifer, or Black Sage, is a species found exclusively in southern California. It is said to have characteristics and properties

SALVIA
Salvia coccinea

similar to those of the other decorative Sages. It, too, is a most attractive addition to a garden.

WILD SAGE, Lyre-Leaved Sage
Labiatae
Salvia lyrata

As one of its common names implies, this herb is distinguished by
its lyre-shaped basal leaves, which are similar in other respects to those
of the other members of the *Salvia* genus. Upper leaves are far apart
from each other, rather sparse, and short-stalked. Wild Sage flowers
from April to June, grows to a height of about two feet, and is found
in sandy soils. It is found from New England south to Florida, and in
parts of the Midwest. The flowers of this showy species are light blue,
somewhat droopy, and trumpet-shaped.

Wild Sage is said to have similar properties to Sage. At least one
herbalist has reported the use of its crushed leaves to make warts
disappear.

WILD SAGE
Salvia lyrata

SAGE
Salvia officinalis

SAGE, Garden Sage
Labiatae
Salvia officinalis

This native of the Mediterranean region is widely cultivated in most parts of the world. It is a small perennial plant, growing to a height of about two feet under ideal conditions. Its spear-shaped leaves are deeply veined and have a rough texture: they are paired on the square stem, and they are somewhat hairy and gray-green in color. The lilac-blue blossoms, which usually appear from June to August, form a spikelike cluster around the topmost part of the stem.

Sage has a strong, distinctive scent and a somewhat biting taste. Once regarded as a valuable medicinal herb, Sage is now relegated to the kitchen for the most part as a uniquely appealing condiment. The genus name, *Salvia,* is said to have been taken from the Latin *salvere,* meaning "to be saved," a possible reference to the herb's elevated status in ancient medicinal lore.

Sage's medicinal uses in the past are innumerable. A tea made from the herb was quite popular to cure oral infections and soothe sore throat. Sage has been included as an ingredient in tooth powders, and it is said to be good to strengthen and heal the gums. The herb has enjoyed some repute as a digestive aid and as a tranquilizing agent for certain nervous disorders. It is thought by some to be a cure for headaches and colds, as well as a great benefit to a failing memory.

In the kitchen, though, Sage has always surpassed itself in practical use. It is used as a seasoning for many meat dishes, especially pork and chicken, and the chopped leaves can be added to soups, salads, and stews in quantities to suit one's taste. The herb has been used as a garnish for fruit salads and as an ingredient in fruit drinks and wine concoctions. It is also recommended to enhance (or disguise) the flavor of organic meats.

Sage is easy to grow in just about any soil, as long as it is in a sunny place. It can be started from seeds sown in the early spring, but it is often more successfully grown from cuttings of the stem set in early summer. The young plants should be thinned to at least a foot apart when they reach a height of around three inches. The leaves can be harvested at any time, and they are dried in the normal way, in a cool, shaded area.

GARDEN CLARY, Clary Sage
Labiatae
Salvia sclarea

This common biennial has typical square stems and it grows to a height of three feet. The leaves of Garden Clary are quite large, furrowed, and somewhat toothed; they appear in pairs on the brownish stems. The plant, which is noticeably aromatic, blossoms into generally pale blue flower clusters that appear on a terminal spike on its upper portion.

Garden Clary was referred to in most Old English herbals. It was once used as an adulterant in wines, and it has been utilized as a substitute ingredient in beer, imparting an increased intoxicative effect.

The herb was once thought to be good for improving vision, but it was more commonly used to aid digestive difficulties. And at least one old herbalist has said that the powdered roots of Garden Clary, if inhaled as a type of snuff, are an effective headache remedy.

While native to southern Europe, Garden Clary is easily cultivated just about anywhere. It is grown from seeds sown in early spring, and the young plants should be thinned to about a foot apart once they have a good start.

GARDEN CLARY
Salvia sclarea

SUMMER SAVORY
Satureia hortensis

SUMMER SAVORY
Labiatae
Satureia hortensis

Summer Savory is a popular seasoning and cooking herb. It is commonly used when cooking beans and other legumes, and especially cabbage and turnips, because it has the ability to diminish the unpleasant smell of those vegetables as well as eliminate the usual flatulence associated with their consumption. Summer Savory grows to a height of eighteen inches under ideal circumstances; it grows from a branched root and has a multi-branched stalk. The short leaves of this hardy annual grow in pairs along the stem, and they are dark green in color. The small, pinkish flowers appear from July to October, and they form small clusters in the upper leaf axils.

Summer Savory is native to southern Europe and the Mediterranean region, but it is extensively cultivated in both Europe and the United States. The herb is highly aromatic: the ancient Romans were said to be fond of its fragrance, and they used it to flavor vinegar for use as a sauce.

The juice of Summer Savory is said to be an excellent pain reliever for bee stings. An infusion of the herb, or tea, is said to be effective for the relief of stomach and intestinal disorders, especially diarrhea. It is also reputed to stimulate the appetite. Some herbalists hold that it makes a great aphrodisiac, also.

In addition to its use in many vegetable dishes, Summer Savory is commonly added to stuffings and meats, including sausages. It is said to enhance the flavor of honey, preserves, and syrups.

Summer Savory is best grown from seeds sown early in the spring. It prefers full sun, but it will adapt to less than exceptional soils. The seedlings should be thinned to a spacing of about nine inches or less, since the plants can support each other more effectively when they're close together. Summer Savory grows well indoors, also, as long as it gets at least five hours of direct sunlight per day. Leaves may be picked at any time, but the herb should be cut for drying just before it is ready to flower for best flavor. The herb may be pulled completely from the ground, or simply cut close, and hung upside down in a dark, well-ventilated room to dry.

WINTER SAVORY
Labiatae
Satureia montana

Winter Savory is similar in appearance, characteristics, and properties to Summer Savory, but the plant itself is slightly smaller and wider. It also differs in that it is a perennial, and the leaves will remain green throughout the winter. The flowers bloom in June, and they are somewhat darker in color than those of Summer Savory.

Winter Savory can be used in the kitchen much the same as the

WINTER SAVORY
Satureia montana

Summer variety. Additionally, at least one old herbalist reported that Winter Savory was at one time dried and powdered and mixed with finely grated bread crumbs to make an appealing breading for meats and fish.

Winter Savory can be cultivated and harvested much the same as its cousin, except that it is difficult to start from seeds, and should be set in the form of root or stem cuttings in the spring. The plant is best divided and replanted every couple of years for better growth.

WILD THYME, Mother-of-Thyme
Labiatae
Thymus serpyllum

Wild Thyme is often called Mother-of-Thyme, since it is thought by some herbalists to be the original species. Its species name, *serpyllum,* is derived from a Greek word meaning "to creep," a word descriptively accurate for this herb. Wild Thyme is a perennial that literally crawls over the ground, forming dense, carpetlike cover, especially in rocky areas. The reddish-brown stems contain numerous branches with paired, bright green leaves and little clusters of purplish flowers which appear from June to September.

Wild Thyme is rarely used in the kitchen like Garden Thyme, but it does possess most of the medicinal properties of its cousin, albeit to

WILD THYME
Thymus serpyllum

an inferior degree. Its forte, rather, is that of a decorative addition to any garden. The carpet of colorful foliage and blossoms created by this herb has been said to be the favorite playground of fairies. And, with absolutely no pun intended, we can relate to you that Wild Thyme tea has been highly recommended as a cure for hangovers.

Wild Thyme will not grow in excessively humid regions, but it can otherwise be cultivated anywhere. It won't do well indoors, however, unless you'd like to carpet your entire living room with the wide-spreading foliage. The plant is great for borders or rock gardens; it is started best from root and stem cuttings. Harvest or dry it in the usual manner.

GARDEN THYME
Labiatae
Thymus vulgaris

This unassuming little herb has myriad medicinal and culinary uses. It will grow to a height of about eight inches, and its round, twiglike stems are branched and contain tiny, paired leaves that are shiny and somewhat oval-shaped. The pale purple flowers, which appear from May to August, grow from the upper leaf axils in small clusters around the stem. Herbalists disagree widely on nomenclature for Thyme. You see, there are said to be several varieties of Thyme, including Garden Thyme. But then, there are several varieties of Garden Thyme, which are determined by the relative size and shape of the leaves. In short, the disagreement arises from the question of whether a given Thyme plant is simply a variety of Thyme, or a variety of Garden Thyme. Over the centuries, one specific Thyme plant may have been given several names, or several Thyme plants may have been given the same name. And so on. Our advice is to take our advice and have a good Thyme.

Garden Thyme is popular in the kitchen, and it has upheld that tradition for many centuries. It was thought by the Romans to be an invigorating herb, and recognition of its antiseptic properties dates from an equally early time. The sweet-smelling plant is said to be highly attractive to bees. Garden Thyme has been an ingredient in herb tobacco, perfumes, and embalming fluids. Distillation of the leaves and flower tops produces oil of Thyme, which has as its most important constituent the phenol Thymol, a powerful antiseptic and disinfectant.

As a result, Garden Thyme has enjoyed numerous medicinal applications related to its aforementioned chemical composition. It was

GARDEN THYME
Thymus vulgaris

often used for various skin conditions, cosmetic applications, mouth-washes and dentifrices, and burns. And internally, Garden Thyme was reputed to be invaluable for the treatment of bronchial congestion, whooping cough, gastric disturbances, and colic.

Garden Thyme has long been thought to compliment the flavor of roasted meat; it is somewhat spicy and sharp-tasting. The herb has been recommended as a seasoning for soups, stuffings, fish, herb butter, meat sauces, and stews. It is also utilized when pickling olives.

Garden Thyme will grow anywhere in the United States, both indoors and outdoors. The plant prefers full sunlight and a light, well-drained soil. It is better grown from root or stem cuttings, since the seeds are extremely slow to germinate. The herb is said to require little or no water: one herbalist has said that Garden Thyme draws its energy from only cosmic sources and that it could seemingly grow from bare rock itself. The herb is harvested at the usual time, just before flowering. If the distillation process is desired, the leaves should be gathered on a dry day after the sun has evaporated the dew from the surface of the foliage.

GERMANDER, Wood Sage
Labiatae
Teucrium canadense

Germander has been used for centuries to heal wounds both inside

GERMANDER
Teucrium canadense

and outside the body. This perennial of the Mint family has lavender flowers that appear in a terminal spiked cluster from June to September. The blossom is small but interesting. It only reaches a length of three-quarters of an inch. The corolla of the flower has two lower lips but looks like it has one. There are five lobes, one of which is elongated and rather flattened. Both the upper lobes and the side lobes of the Germander are short. Four stamens are present. The stem is covered with a downy hair on each of its four surfaces. That same dense fiber covers the bottom of its long, toothed, lance-shaped leaf. The leaves appear opposite on the stem. Germander thrives in the thickets and woods, as well as on the shores of fresh streams and ponds.

Germander is found naturally in southern Canada and throughout the United States.

The common name Germander stems from a Greek word for Ground Oak, *chamaidrys*. A species of Germander was at one time used as a substitute for hops in England. The herb makes a good hedge, as it grows low to the ground.

Germander was commonly used to heal cuts in the skin; it was especially helpful for scab formation. A vapor of the plant, when inhaled, is said to soothe the pain from throat irritation. It is said to be a diuretic and a tonic. Colds and fevers, tuberculosis, and kidney and bladder trouble are eased by ingesting an infusion of Germander.

It takes about a month for the seeds to germinate, so it is often recommended that root division in the fall or spring cuttings be used to propagate the species. A rich, well-drained soil in a sunny location is preferred. Remember, this herb grows to only a foot and a half in height and can be neatly trimmed to form a border hedge.

LAURACEAE
The Laurel Family

The family consists of mainly aromatic woody plants with alternate leaves that are simple in structure. The flowers are bisexual or unisexual, and symmetrical. They are usually yellow or green in color. Botanists agree on about 40 genera and approximately 1,000 species, occurring mostly in the tropics, but with a few examples found in North America.

BAY
Lauraceae
Laurus nobilis

Bay is a well known cooking herb used for its delightful seasoning abilities. The plant is a shrub that may attain a height of ten feet in the United States, where it is found only under cultivation. It is a perennial evergreen, with dark green, shiny, smooth leaves. The herb rarely blooms in this country, but in its native Mediterranean soil it produces tiny yellow blossoms, followed by purplish-black berries.

BAY
Laurus nobilis

The Greeks and Romans were said to have made crown wreaths from this herb to adorn their heroes, poets, and athletes. Some herbalists have speculated that the term "bachelor," referring to a college degree, is actually derived from the Latin *bacca-laureus,* meaning "Laurel berry."

Bay leaves are commonly used during cooking to flavor most fish, meat, and vegetable dishes, and especially stews.

The leaves and berries are regarded as somewhat narcotic, and the berries of Bay have been used in the past to produce spontaneous abortions. An oil distilled from the leaves has been employed to relieve earache.

Bay is usually cultivated from a starter plant, either indoors or outside. It requires a moderate amount of sunshine and a well-drained, moist soil. The leaves may be gathered anytime and used either fresh or dried.

LEGUMINOSAE, or FABACEAE
The Bean Family or The Pea Family

With a membership of about 500 genera and over 15,000 species, the Pea family qualifies as one of the largest and most significant in terms of sheer numbers. The family also has great economic significance, in that its members comprise an extensive food source for man, both directly and indirectly. Man consumes many of the herbs found in the Pea family, including Peas, Peanuts, and Soybeans; and man's livestock and domestic animals consume Alfalfa, Clover, and Field Peas. Also, the plants in this group are well known for their ability to regenerate soil fertility. Not all of the family members are friendly, however. Some species, known as loco weeds, found commonly in the West, are poisonous. Still other plants here are grown as ornamentals. There are three different types of flowers among the members of the Pea family: the Pealike flower, which resembles the prow of a boat; the Acacias, which are radially symmetrical; and the Sennas, which are bilaterally symmetrical. The leaves are pinnately or palmately compound, and sometimes there are no leaflets at all. The fruits are usually one-chambered pods, with one or two seam openings.

WILD SENNA, American Senna, Maryland Senna
Leguminosae
Cassia marilandica

Wild Senna is a perennial plant that will reach a height of four to six feet under ideal conditions. It prefers the rich soils of the eastern United States. The herb has a round stem that is somewhat hairy. Each leaf has from eight to ten small, narrow, pointed leaflets. The flower is bright yellow; it forms clusters of gold borne along racemes. They appear from June to September and are followed by four-inch-long flat pods that are leguminous. The purplish-brown anthers of the flowers make a striking contrast with the yellow petals. Wild Senna is found in southern Canada and all over the United States.

This herb was formerly called Maryland Senna because it was commonly found there. The seeds have a somewhat soapy taste but have been used by some as a breath freshener. The leaves are gathered when the plant is in bloom.

The Shakers were said to have formed the leaves into oblong cakes

WILD SENNA
Cassia marilandica

like they did other herbal preparations. Wild Senna has long been known for its proclivity to act on the bowel. It has a tendency to cause pain in that area, however, and should be used in combination with other herbs. It is said to be an effective laxative, diuretic, and vermifuge. You may prefer simply to have the herb around for its attractive blossoms, though.

Wild Senna makes a good back border in your herb garden, because of its height. Also, it may provide some shade for the other herbs in your collection. Wild Senna likes full sun or just a touch of shade and a well-drained soil. It often grows wild in moist, open woods, and disturbed areas. You can take the dried seed pods and place them in the ground where you want them to grow in the fall of the year. Give the seedlings plenty of growing room and watch your Wild Senna take off.

LICORICE
Leguminosae
Glycyrrhiza glabra

Licorice is a small perennial that is a native to parts of Europe and Asia. It is widely cultivated, however, for its precious rootstock, from which the well-known extract is derived. The rootstock is brown externally, with a woody texture: the taproot may reach a depth of three to four feet. Inside, the rootstock is yellow in color and has a sweet, unique taste. The plant has pinnate leaflets that are accompanied by racemes of usually purple flowers, followed by smooth seedpods. The flowers appear from June to August.

Licorice has a long history of use. Although the Greeks were the first to record the use of the herb, it is quite probable that its properties were known long before then. The Greek herbalist Dioscorides is said to have named the plant with two Greek words: *glukos,* meaning "sweet," and *riza,* meaning "root." Hence, the genus name, *Glycyrrhiza.*

Licorice is best known for its use as both a cough remedy itself and a flavoring to disguise the taste of other bad-tasting cough medicines. Licorice is sold commercially as a stick, a paste, a powder, and a fluid extract. All of the above come from the root.

Licorice is often used by brewers in the making of various ales, and it has been utilized in the manufacture of certain chewing tobaccos.

Licorice may be planted in late fall, but early spring is the preferred time generally. It is best started from pieces of the root or runners thereof that have obvious eyes, or buds. The cuttings should be placed about four inches below the surface, in widely spaced rows. The space between the rows is not for the benefit of the plant, but for the grower, who must have room to dig deeply along the mature plants to gather the valuable root. Licorice prefers a rather consistently warm climate, a sandy soil, and plenty of moisture during the growing period.

MELILOT, White Sweet Clover
Leguminosae
Melilotus alba

Melilot is a tall plant that has a vanillalike flavor when dried. It grows from two to four feet high and is found in dry fields all over the United States and Canada. The stem is smooth and erect with many branches. The leaves are narrow and oblong. Growing from alternate sides of the stem, the leaves are trifoliate and smooth. Flowers appear along thin racemes; they are tiny and white or yellow in color. The keel of the flower is shorter than the other parts and has a lot of honey in it. The flower's aroma comes from a chemical, Coumarin, which is also found in Fenugreek, Woodruff, and new mown hay.

The genus name, roughly translated, means "honey lotus," and reflects its popularity among the honey-loving insects. Melilot is a

MELILOT
Melilotus alba

valuable forage crop, and it adds nitrogen to the soil through its root nodules. It is believed to be a native of western Asia. Melilot is sometimes used in the making of herbal snuffs and smoking mixtures. It makes a fragrant filler for herb pillows and potpourris. Melilot is grown in the plains states, where Kansas is the leading seed producer. Out of the million acres planted in this country, 12,000 tons of seeds have been produced.

The leaves of Melilot contain specific anticlotting agents and an extract from the leaves of this herb was said to have been used to treat President Eisenhower after a severe heart attack. Large doses of the herb have been known to cause vomiting and other symptoms of poisoning. Nevertheless, Melilot has been used as an antispasmodic. It is said to rid the body of excess water. When made into a salve, it can be used externally to reduce swelling and help ease skin lesions. The herb is also reputed to ease the pain of aching joints and headaches; it has been made into an eyewash to ease inflammation. A tea made of Melilot can soothe stomach pains and coughs.

Melilot is used to flavor cheeses. The flowers may also be tossed into salads.

Melilot can be grown in the spring from seed, and if left to mature, it will propagate itself freely in the fall. The herb will adjust to most soils and likes a lot of sun. The flowering plant is a good source of nectar for honey-producing bees.

Melilotus officinalis is a yellow-flowered species that can be used interchangeably with White Sweet Clover, as Melilot is otherwise known.

ALFALFA
Leguminosae
Medicago sativa

The violet-purple flowers of Alfalfa are well designed for cross-pollination by bees. When the "keel" of the flower, or two bottom petals, is pressed downward, the flower is said to be "tripped," meaning that the stigma and stamens are released and have snapped upward allowing pollen to be received. The leaflets of Alfalfa are ovate and oblong, and they grow from an erect, smooth stem. The taproot is characteristically long. The fruit is a spirally coiled seedpod that contains several seeds that follow the late summer bloom.

Alfalfa is widely cultivated for domestic animal forage. It contains many vitamins and organic minerals, and it is one of the richest sources of organic salts.

Alfalfa makes a pleasant tea, especially when combined with a little mint. But the tea also has medicinal value. It can stimulate the appetite and pep one up. You can take it every day as a tonic, reportedly, and some say that it will relieve urinary and bowel problems

ALFALFA
Medicago sativa

and cure peptic ulcers. Combined with a little lemon grass, Alfalfa tea helps to eliminate body water.

Alfalfa thrives in wide open soil that is well drained in most places on the Continent. So if you can use Alfalfa in your garden, keep in mind that it can grow up to eighteen inches in height and will therefore need a corresponding amount of space. Sow from seeds in the spring where you want the plant to grow.

Alfalfa is said to be an aid to alcoholics and narcotic addicts who are in the process of detoxification. It was thought to help rebuild decayed teeth if eaten regularly. Alfalfa sprouts contain more protein than wheat or corn, and the herb has long been a symbol of life.

RED CLOVER
Leguminosae
Trifolium pratense

Red Clover is a common cover crop that is easily distinguished by its tight clusters of pink-purple flowers and honeylike fragrance. The plant is a perennial that grows to a height of about two feet on meadow lands, lawns, and along roadsides. It has the characteristic three leaflets that Clover is famous for. Red Clover blooms from May to September.

Red Clover is one of the most widely used plants for crop rotation,

RED CLOVER
Trifolium pratense

since its roots store nitrogen in small nodules, thereby improving soil fertility. It is an excellent forage crop. The herb is said to be a native of Asia, but it was widely distributed throughout Europe during the Middle Ages. The plant is said to symbolize industry, since it is often allied with the thought of the "busy bees." Red Clover is high in protein, phosphorus, and calcium.

The herb has had many medicinal uses in the past. A fluid extract has been employed as an antispasmodic. A poultice of Red Clover foliage was once applied to cancerous growths and to athlete's foot. The ground blossoms of Red Clover were said to have been used in the preparation of a cigarette that, when smoked, was thought to be an asthma preventative.

Red Clover is easily grown from seeds, but it prefers a moist, sandy soil for best results.

WHITE CLOVER
Leguminosae
Trifolium repens

This is the common lawn plant that one investigates in search of the famous "four-leaf Clover." Like Red Clover, it is a perennial, but White Clover is a creeping plant from which white flower clusters and short leaf stalks arise. The mature plant reaches a height of only ten inches. It flowers from May to October, and it is commonly found in lawns, along roadsides, and some fields.

FENUGREEK, Bird's Foot
Leguminosae
Trigonella foenum-graecum

Fenugreek is an erect, annual herb about two feet high. The leaves appear in threes; they are toothed. Small brownish oblong seeds that have a deep furrow and are contained in a long, narrow sicklelike pod. The stem is hollow. The off-white flowers, which grow to about an inch long, form at the joint of the leaf and the stem. The flowers turn into six-inch-long pointed seedpods. When it is very young, the plant looks somewhat like its cousin, Sweet Clover. The flowers are very fragrant.

The seeds have been highly valued for centuries. They are said to

FENUGREEK
Trigonella foenum-graecum

have a flavor that resembles Celery and were highly regarded by Egyptians, Greeks, and Romans. The species name, *foenum-graecum,* means "Greek hay." Fenugreek was used to scent inferior hay.

The seeds have been ground to give a Maple flavoring to confections. Fenugreek is also one of the ingredients of Curry.

Fenugreek seed tea has been used medicinally as a gargle to soothe the throat. A mucilage made by soaking the seeds until they swell into a thick paste was said to be equal to Quinine in its ability to quell fevers. The paste has also been used to soothe the stomach and assist in the treatment of diabetes. Externally, the seeds in the form of a decoction have been used to heal lesions and absesses on the skin.

Fenugreek thrives in well-drained soil in full sun. Cover the seeds in the spring with a thin layer of soil; the young seedlings should be thinned to about four inches apart. It will take about four months to go to seed; and when harvesting Fenugreek, pull the entire plant from the ground and turn it upside down to hang in a dry place out of the sun.

LILIACEAE
The Lily Family

The herbs of the Liliaceae family include some valuable food sources, such as Garlic, Asparagus, Onion, and Chives. The plants are mostly perennials, and they often grow from bulbs. Their leaves are usually simple and slender and are alternate on the stem. Flowers of the Lily family are sometimes quite attractive; they're symmetrical and usually bisexual. The fruit is a capsule formed on three sections, or it can sometimes be a berry. Speculatively, there are about 250 genera and perhaps as many as 6,000 separate species.

STARGRASS, Colicroot
Liliaceae
Aletris farinosa

The white flowers of Stargrass appear on a long, spikelike stem. The stem is round. The flower petals and sepals are fused to form a toothed, tubelike blossom that is somewhat granular on the outside. The long, pointed leaves are arranged at the very base of the stem:

STARGRASS
Aletris farinosa

some leaves do appear further up on the stem, but they are very small. The fruit is a triangular capsule. Stargrass flowers from May to August, and the fruit appears shortly thereafter. Stargrass grows in sandy, wooded locales in its native North America. The tuberous root grows almost parallel to the surface of the ground and it has hair fibres that extend from its lower side. It is these roots that were gathered extensively before the 19th century for their supposed value in the treatment of colic.

The starchy root is very bitter in taste; its tonic properties are extracted with alcohol. Stargrass root has been used in the treatment of hysteria and to tone up the stomach muscles. Precautionary measures must be undertaken because the root does have some narcotic side effects.

Stargrass grows well in acid soils that are rich in humus and well shaded.

WILD GARLIC
Liliaceae
Allium canadense

Wild Garlic leaves reach a height of a foot and a half in low woods and meadows. Stemming from the oval-shaped bulbet of a root, the

WILD GARLIC
Allium canadense

grasslike leaves conceal the smooth support of a pink or white rounded cluster of flowers. The leaves are long, linear, and flat. The flowers growing about a half inch wide form three petallike sepals and the same number of petals. Three modified leaves appear beneath the flower cluster. The flowers may later be replaced by small oval-shaped bulbs, some with long taillike extensions. May, June, and July are the months this onion-scented plant chooses to bloom from Ontario and New Brunswick south to Florida and west.

Wild Garlic is a native to North American shores and is often harvested for its bulb, which has an oniony flavor. Milk and dairy products tend to carry the flavor of this herb if cows have grazed upon it extensively.

ONION
Liliaceae
Allium cepa

A hollow blue-green stem rising over a foot in height signals this often-cultivated biennial herb. Most people are familiar with the bulb of the Onion, which has characteristic parchmentlike sheathing to protect it. The leaves spring from this bulb, and they are shorter than the stem. The leaves match the bluish-green stem in color and they have the same hollow construction. A cluster of greenish white flowers blooms

ONION
Allium cepa

in summer from June to August, followed later by tiny bulbs.

Finely ground onion soaked in gin is given as a cure for gravel and water retention. Onion is believed to be a cure for baldness, nose and throat infections, and insomnia when eaten raw late in the evening. Onion syrup is prescribed in farm kitchens for coughs and the flu. In the Middle Ages, Onions mixed with honey, salt, and chicken fat were applied to the skin to remove blemishes. One source says honey poured over thin slices of raw Onion that has remained undisturbed overnight can be sipped slowly for asthma relief. Children were encouraged to eat a large number of Onions to stop heartburn and aid digestion and to kill worms.

The finest restaurants offer Onion soup as a preliminary to both the most sumptuous meal and the most simple sandwich.

Onions can be grown in the spring of the year from small bulbs planted from four to six inches apart in plenty of sunlight. Bend the stalk to increase the size of the bulb and keep the dirt away from the top of it as it grows. Air the Onion bulbs in a cool, dry place for several weeks and they should keep for you for several months thereafter.

EGYPTIAN ONION, Top Onion
Liliaceae
Allium cepa, var. aggregatum

This variety of Onion gets its name from its supposed country of origin. It is valued today for its mild-flavored bulb. Egyptian Onion is unusual in that it forms a cluster of reddish-brown bulblets at the top of the tubular stem. The stem, or scape, grows to a height of three feet. It "nods" to the ground as the bulblets mature, allowing them to take root. Meanwhile, underground, little activity is seen in the growth of that bulb.

The flavor of the Egyptian Onion is more delicate than other Onions, but provides a delightful pick-me-up in salads and with potherbs. You may also find it in pickled form in your martini. The tubular leaves, when young, may also be used as salad greens. Any of the Onion greens may be frozen for later use.

Egyptian Onion prefers a nearly neutral soil. New plants can be started from bulblets which have been dried in a cool, dark place from last season. Each bulblet can be covered with a half-inch of soil and placed four to eight inches apart. When the dark blue-green leaves begin to wither, dig them out and hang them in a cool, dry place for use throughout the winter. Chop off the stem when the bulblets first

EGYPTIAN ONION
Allium cepa, var. aggregatum

appear to extend the growth of the leaves. Save the bulblets for fall or spring planting.

GARLIC
Liliaceae
Allium sativum

Common Garlic grows to a height of three feet. The leaves are similar in appearance to grass in that they are long, narrow, and flat. They grow from cylindrical sheaths surrounding the stem. The most important part of this plant, however, is the bulb, which consists of a group of cloves bound together by a parchmentlike skin. The bloom is an umbel of small white flowers which number up to thirty. They emerge from a hooded sheath which eventually falls away.

Garlic has long been associated with the common man, and despite its offensive odor, it is much sought after in the kitchen as well as the medicine chest. Garlic was mentioned frequently in the Old Testament, and even 3,000 years before Christ, it was used by the Hebrews in ceremonial dishes. Garlic was often hung around the necks of cold sufferers and was said to keep away evil spirits, including Dracula. One of the best-known herbs, Garlic was said to be a healer and a strengthener. Egyptians consumed it while building the pyramids, and

GARLIC
Allium sativum

Roman soldiers were said to have eaten Garlic while on long marches between battles.

Garlic adds a distinctive touch to just about any food. It becomes overbearing, however, if you don't use it properly. But taste for the herb is a relative matter, so you'll have to experiment with it to determine your tolerance level. Garlic is popular with fish, meat, and vegetable dishes. Any lover of escargot would be lost without the haunting taste of Garlic. If you'd rather not have the taste or aroma of Garlic lingering with you, though, try chewing on a little fresh Parsley to aid in the abatement of the odor.

The strong odor of Garlic comes from the essential oil, Allyl, which is present in all members of the *Allium* genus. Garlic is said to stimulate the digestive system, but it will also soothe an upset stomach, supposedly. It has been recommended for treatment of high blood pressure. The juice of Garlic has been used in the past to treat wounded soldiers, assist in the control of various skin diseases, and ward off the plague and other ills. Even today, everyone knows that it's a good idea to eat a Garlic sandwich to ward off a cold. Garlic has been used to expel worms in the intestinal tract, and it is said to be especially efficient for the expulsion of pinworms. A Garlic-based cough syrup, tempered with other herbal sweeteners, is said to be useful to rid one of coughs and hoarseness.

Garlic is pretty easy to grow: it thrives in a slightly acidic soil.

Each single clove of the bulb can become a plant when set two inches under the soil and about half a foot apart. During July and August the leaves will turn yellow; this is your signal that you should pull the plant from the earth for drying out of the sun's harsh rays. You can harvest Garlic a little faster if you retard the growth of the plant by pushing the top down as the yellowing phase begins. Garlic needs a sunny spot, but keep the ground moist.

CHIVES
Liliaceae
Allium schoenoprasum

Chives is found as a cluster of hollow grasslike leaves that grow to a height of about one foot. The lavender flower looks like a fluffy ball of about one inch in diameter atop a long, thin stalk; the flowers bloom late in the spring. It is a hardy perennial.

Little is known about the past uses of Chives other than the belief that it was cultivated in ancient China. It is a valuable and tasty seasoning and a decorative border plant for herb gardens. Chives is also a great indoor herb.

Chives is best planted in a moist soil; it likes a sunny area. Seeds

CHIVES
Allium schoenoprasum

may be sown in the spring. The seeds should be covered with about one-quarter inch of soil. Seedlings need not be thinned for the first year's growth, so that the small clumps of them may be transplanted the following year. It's best to have at least five or six bulbs in each transplanted clump. And leave about six inches of space surrounding each cluster.

WILD LEEK, Ramp
Liliaceae
Allium tricoccum

Arising from an onionlike bulb to a foot and a half in height, Wild Leek leaves die back when the herb flowers, leaving an umbel of creamy white blossoms atop a smooth stem. The flowers have three petals which resemble cups and three sepals: they are the same color. The clusters are about one and one-half inches wide, and they appear from June to July in rich, moist woods. The flowers are followed by a three-chambered fruit. Wild Leek grows naturally from Canada's provinces of New Brunswick and Nova Scotia and Quebec south through New England to Georgia and west.

Native to the North American landscape, and seldom found anywhere else, Wild Leek has long been a herald of spring; and in late

WILD LEEK
Allium tricoccum

April people of the Great Smoky Mountains as well as Amish folk everywhere delight in harvesting the early bloomer for Wild Leek parties known as "Ramp Festivals." Great steaming caldrons of Wild Leek soup bring together neighbors separated through the long, dark winter months. Wild Garlic is the only member of the Allium genus that dares to bloom before Wild Leek.

Both the leaves and the mild onion-flavored root are used in soups and salads. Juice from the Wild Leek was used by the American Indians to treat insect bites and beestings. Look for its distinctive broad leaf.

Plant the bulbs of Wild Leek in the spring in rich, moist soil. Make sure they're kept well watered. Harvest just before flowering for the most delicate flavor.

ASPARAGUS, Spear Grass
Liliaceae
Asparagus officinalis

The young shoots of Asparagus are quite familiar as a common vegetable. If left to mature, however, those shoots can spring up to a height of five feet or more. The herb is a perennial, and it grows from long, thick roots. The needle-shaped leaves are somewhat inconspicuous, appearing to be clustered in scales on the stem and branches.

ASPARAGUS
Asparagus officinalis

The flowers, which bloom in May and June, are greenish-white; they're quite small and bell- or trumpet-shaped. The fruit, which ripens in August, contains within its red berry exterior black seeds that are about a third of an inch in diameter.

Asparagus roots and shoots contain the chemical Aspergin, which is said to stimulate the kidneys and liver when ingested. The herb should not be eaten, therefore, when one is suffering from inflamed or irritated kidneys.

The young shoots of Asparagus can be eaten raw or cooked. They are best cooked in very little water for a short time. The roasted seeds from the red berries have long been used as a coffee substitute, especially in Europe.

Medicinally, Asparagus has been used as a diuretic and a laxative. It is said to induce sweating and increase urine production, and it is thought to be a good fibre source.

Asparagus is often found as a garden escape in waste places just about anywhere. It is easily grown from seeds sown in the spring and is a common garden companion to Parsley and Tomatoes.

LILY-OF-THE-VALLEY
Liliaceae
Convallaria majalis

Lily-of-the-Valley is a fragrant and comforting plant to have around the house. Two basal leaves rise from the rhizome: they are dark and green, and one is often longer than the other. The leaves act as a rain conductor to the rootstock because of their concave surfaces. The stem rises from a scaly sheath to a height of about a foot. There are no leaves on the stem, and white bell-shaped flowers nod downward on one side of the top of the scape or stalk. Lily-of-the-Valley flowers from May to June. The flowers later develop into red berries that are rather fleshy on the outside but hard on the inside.

The herb is found throughout North America. It is a native of Eurasia, but has become a common garden escape in the United States and Canada. The species name, *majalis,* means "that which belongs to May." An Old English legend states that deep in a certain wood, Lilies-of-the-Valley sprang up where drops of blood had fallen from a saint who was engaged in mortal combat with a heinous dragon.

Lily-of-the-Valley has been used as a substitute for Digitalis, especially in cases where the patient has developed a tolerance for the former. But care must be used because the herb contains two glycosides

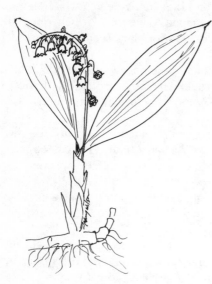

LILY-OF-THE-VALLEY
Convallaria majalis

that can cause upset stomach or an irregular heartbeat.

Lily-of-the-Valley is pretty easy to grow since its creeping underground stem sends new shoots up in the spring. A light porous soil in the shade is best but the fruit may also be used to propagate the species in the spring. Transplant about every four to five years in the fall. Use roots that are six inches long and about six inches apart; make sure the crowns are at the top of the deeply planted root.

WHITE DOG-TOOTH VIOLET
Liliaceae
Erythronium albidum

Like its close cousin, Adder's Tongue, White Dog-Tooth Violet, has mottled leaves and bell-shaped flowers; but this variety has narrower leaves and a white flower that is lavender on the outside instead of yellow. It's found throughout the continent in sparse woodland that has rich soil. It is also an early spring bloomer.

ADDER'S TONGUE, Trout Lily, Dog Tooth Violet
Liliaceae
Erythronium americanum

Adder's Tongue is a perennial plant that commonly signals the

ADDER'S TONGUE
Erythronium americanum

advent of spring. Its striking yellow drooping flower appears in quantity in March and fades away by early June. The herb is distinguished by a pair of brown-spotted leaves that spring from the base of the ten-inch-long stem. The stem of this native plant springs from a deeply rooted bulb. The bulb is brown on the outside and a brilliant white on the inside. The flower petals curve backward noticeably and are often tinged with purple. Adder's Tongue is very striking in the spring in rich wooded areas where it can sometimes be found in large colonies which spread like a yellow blanket over the moist spring earth. It is found throughout the United States.

Even though Adder's Tongue is commonly called Dog-Tooth Violet, one must keep in mind that is is a member of the Lily family. The "Dog-Tooth" reference is made because of the shape and color of the underground root. The other common name, Trout Lily, points out the fact that there is a similarity between the markings on the leaves and mottled trout bodies.

Some herbalists have gathered the leaves in the spring when they first appear as an addition to salads. The raw leaves are said to be juicy with a sweetish cucumberlike flavor.

An infusion of the leaves and roots has been used medicinally to draw out infections and to soothe open wounds and diseases of the skin. The infusion mixed with a little cider is said to go down a little easier. For skin problems, a poultice of the leaves applied directly has been used with some success.

When attempting to cultivate Adder's Tongue, try to simulate its natural growing conditions as closely as possible. Once again, it prefers rich soil with lots of humus in a canopy of shade. You may not have much luck in transferring the wild variety, but if you dig the plant after its flower and set of leaves have died away, you may be successful if you are sure to place the bulb at least three inches under the ground. Some varieties may propagate by sending out rootstocks.

MINNESOTA ADDER'S TONGUE
Liliaceae
Erythronium propullans

Like the name implies, this herb is a native of Minnesota and is rarely found outside the north central state. Minnesota Adder's Tongue appears in April through May sporting a pink flower instead of the yellow flower found on the Adder's Tongue or Dog-Tooth Violet. The Minnesota variety forms a bulblike growth partway up the stem of the flower. Look for this herb beneath a bough of trees where it likes to grow.

INDIAN CUCUMBER ROOT
Liliaceae
Medeola virginiana

Nestled deep in the rich woods, Indian Cucumber Root forms several greenish-yellow flowers from the center of three leaves. These leaves form at the top of a slender wooly stem: from those leaves grows a nodding stalk that can be so weighted by the flowers that form atop it, that it may in fact bend down below the level of the leaves themselves. The flowers have three curly petals and three curly sepals that look like petals, and three brownish curling stigmas in the ovary. The stem has two sets of lance-shaped leaves. The larger group is partway up on the stem, and the smaller group of one to three leaves is at the top. Indian Cucumber Root grows to a height of two and one-half feet in moist woodlands all over the continent from May to June. The flower is replaced by a dark purple berry. The root can grow up to three inches long and an inch in diameter. It is white and brittle and tastes like a Cucumber; hence, the name.

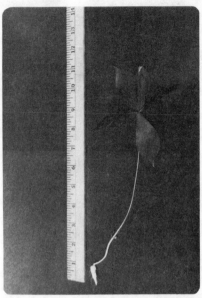

INDIAN CUCUMBER ROOT
Medeola virginiana

Native Americans used the root as an important food source. The herb has become somewhat scarce in recent times, so only dig it from an area that is well populated.

To propagate Indian Cucumber Root, dig the root after the flower has faded. Place it in a rich, loamy soil in a shaded area toward the fall of the year. Make sure the stem of the plant is severed just above the soil line. This will allow all the plant's energy to remain concentrated in the root in preparation for a spring growth spurt. The root must be kept moist throughout the growing season.

SOLOMON'S SEAL
Liliaceae
Polygonatum biflorum

Appearing like an arching green feather in moist thickets, Solomon's Seal bears greenish, bell-shaped flowers in pairs of two. These flowers hang from the leaf axils and appear from May to June, followed by a dark blue berry. The six-lobed, six-stamened flowers reach a length of two-thirds of an inch. The leaves are light green and smooth on both sides, with parallel veins. They appear in pairs, decreasing in size near the tip. Solomon's Seal can be found naturally on the Northern Continent.

SOLOMON'S SEAL
Polygonatum biflorum

There is a long and colorful history associated with this herb, from which the plant gets its name. When the flower is dipped in ink before pressing on paper, it forms a six-pointed star like the Jewish Star of David. That star is also known as the Seal of Solomon. One ancient legend says that King Solomon was able to quarry some particularly tough rock when bearing this plant with him. Some herbalists, however, say that the name was assigned to the herb because it was used to seal wounds. Yet another theory claims the name comes from the shape of the scars on the rootstock, which resemble Hebrew script.

Found commonly in the woods, one must often brush aside the leaves of Solomon's Seal to see its nodding whitish flowers. The starchy root was gathered by the American Indians and later by the colonists, as a food supplement. The root grows horizontally as a thick rhizome. Native Americans also brewed a tea to ease internal pains from this herb. A poultice of the root is said to heal blemishes and skin problems and to counteract the blisters of Poison Ivy.

Solomon's Seal can be started from seed in the late summer or early spring. Sow the seed where you want the plant to grow. Solomon's Seal will do well in shade or partial shade, in a loamy soil. The root may be transplanted in the late summer or early fall. Keep the root moist until it is planted, about three inches under the soil surface. Keep the new sprout well watered.

SARSAPARILLA, Jamaica Sarsaparilla
Liliaceae
Smilax ornata

Sarsaparilla is a large, prickly stemmed perennial that has a long, spreading root. It is a climbing plant with large, veined leaves that are stalked and appear in an alternating pattern on the stem. Sarsaparilla root has an orangish tint and a distinctive bitter taste. It is native to Central America.

Sarsaparilla takes its name from a combination of two Spanish words which refer to the fact that the herb is a rather briary vine. It was once thought to be a remedy for syphilis, and it has been used to treat gout, rheumatism, and other chronic diseases. A decoction made from the root was once thought to be a good springtime tonic, and it may have been the base of a once popular western soft drink.

HELLEBORE, American Hellebore, Indian Poke
Liliaceae
Veratrum viride

Hellebore is a large plant with huge, parallel-veined leaves that seem to surround the stem at their bases. Clusters of small, prickly, greenish-yellow flowers form at the top of this herb from May to July. This two to seven foot giant prefers a wet habitat, especially swampy areas. Hellebore grows naturally in southeastern Canada and central and northeastern United States regions. The fruit is a three-chambered capsule. The root, from which the medicinal properties of the herb are extracted, is dark brown in color outside, and a whitish color inside.

Hellebore is a potentially deadly poison. Both its root and foliage contain constituents that will cause severe intestinal irritation and unpredictable heart reaction.

Hellebore has been used in the past to make an ointment that is said to be effective in controlling itching. It has also been used to destroy parasites; especially in the hairy parts of the body. The alkaloids contained in Hellebore are known to have the capability to destroy white blood cells and to slow the rate of heartbeat. Violent vomiting and incredible nausea are the results if Hellebore is ingested. Medical assistance should be sought immediately if Hellebore poisoning is suspected.

SPANISH BAYONET
Liliaceae
Yucca aloifolia

This variety of the *Yucca* genus has toothed hairless leaf margins
that distinguish it from *Yucca filamentosa,* or Yucca.

SOAPWEED
Liliaceae
Yucca glauca

Soapweed is a western variety of Yucca. It has marginal hairs
along its rigid, bayonet-shaped leaves. Its stalk can reach a height of
four feet; it's found as far east as Iowa and Missouri. The fruits, which
form oblong capsules, can be cooked and eaten with the seeds removed.
The petals of the flowers are used in salads, and the roots are formed
into a soap substitute.

YUCCA, Bear-Grass
Liliaceae
Yucca filamentosa

Striking white flowers on a stem reaching a height of ten feet set
this plant apart in the garden. The nodding bell-shaped flowers perch
atop the thick stem, which has its origin in a clump of rigid lance-
shaped leaves. The leaves have loose threads along the margins and

YUCCA
Yucca filamentosa

they can be nearly three feet long and about two and one-half inches wide. They taper down at the ends like the point of a dagger. The flowers bloom from June to September, sporting three petals, three petallike sepals, and six stamens. Yucca grows most commonly along sandy beaches, sand dunes, and in old fields from New Jersey to Florida and along the Gulf Coast.

The root of Yucca is said to be edible, but it was also used by the Indians as a form of soap to wash clothes. Additionally, the leaves of the herb were at one time soaked, beaten, an woven into bedding material.

MALVACEAE
The Mallow Family

The members of this family of herbs are usually velvety or hairy with single flowers or branched clusters of flowers. The flowers are usually bisexual, have three to five sepals that are partly united, and have five petals that are separate. You'll find that many stamens are joined together at the stalk and form a tube; all these parts are connected at the base of the ovary. The simple leaves appear alternately on the stalk, often with veins that appear palmately. The leaves are lobed or deeply divided. The fruit will be at least five-chambered: the chambers will separate from each other. The fruit can also be in the form of a capsule or berry. There are 85 genera and nearly 1,500 species.

VELVETLEAF, Pie-maker
Malvaceae
Abutilon theophrasti

Velvetleaf acquired its alternately popular name, "Pie maker," because of the appearance of its fruit, which is roundish with crimped edges, similar to those of a pie crust. The plant grows to a height of six feet, and it has large, heart-shaped leaves which grow to a length of eight inches. As with the other members of the Mallow family, the leaves are somewhat velvety. The flowers, however, are smaller, as with the Cheese plant. The flowers are five-petaled, with numerous

VELVETLEAF
Abutilon theophrasti

stamens that form a tubelike structure. The yellow flowers appear from July to October. Although a native of India, Velvetleaf is commonly found everywhere in eastern North America, with the exception of the southeast portion of the United States.

The large, hairy leaves of Velvetleaf have been used as a smoking tobacco substitute, especially for use with pipes. It is said to have a soothing effect on the throat. Also, the seeds of Velvetleaf have been ground up for use as a feed for wildlife, including their addition in different types of milk substitutes for animals. The leaves have also been known to become somewhat aromatic when dried, and therefore, have been used for that purpose.

MARSHMALLOW, Sweet Weed, Mortification Root
Malvaceae
Althaea officinalis

Marshmallow is a plant of approximately four feet in height; it has only a few lateral branches, usually. The thick leaves can be round, oval, or from three- to five-lobed. Most often, they are about three inches long and about half as wide. The flowers are somewhat trumpet-shaped, five-lobed, and are a pale purplish pink color. Marshmallow

MARSHMALLOW
Althaea officinalis

flowers from August to September, followed by a flat, round fruit commonly called "cheeses."

The entire Marshmallow plant contains a highly emollient mucilage. Its family name, Malvaceae, is derived from the Greek, *malake,* meaning "soft," and its genus name, *Althaea,* is from the Greek *altho,* meaning "to cure." It was generally thought to be the source of excellent remedies for softening and healing both inwardly and outwardly.

Marshmallow was once used as a reliable and pleasantly soothing sore throat and cough remedy. It also gained a reputation as a good internal treatment for ulcerated stomach or intestinal lining. And a lotion made from the plant mucilage has been used as a hand lotion to soothe dry or chapped skin.

The roots of the plant are said to be good to eat, if sliced crosswise, parboiled, and then fried in butter or oil with onions and seasonings. The water in which the roots were boiled, in turn, can then be beaten like egg whites into a froth to make such things as chiffon pies. A dish of Marshmallow was said to be a favorite delicacy of the Romans. And in times of famine or crop failure, the plant was often collected for food by the poorer inhabitants of Syria. The leaves and young tops of the Marshmallow are used by the French to make a spring salad.

Marshmallow has been successfully cultivated since ancient times. Seeds are best sown in the spring, or carefully divided offsets of the roots, gathered in the fall, will grow well. The plant does best if spaced about two feet apart in moist soil. Whether from cultivated plants, or those found in the wild, the leaves are best picked in August, when the flowers are just coming into bloom. They should be picked in the morning after the dew has been evaporated by the sun.

HIBISCUS, Rose Mallow
Malvaceae
Hibiscus coccineus

The Hibiscus is a rather large plant, attaining a height of from six to ten feet. Its flowers are exceedingly striking. They are large—from six to eight inches in diameter—and deep red in color, with five petals. Numerous yellow stamens form a tube around the extended style. The leaves which appear alternately on the stem, are somewhat divided into coarsely toothed, pointed segments, similar in appearance to a maple leaf. The plant flowers from June to September.

This species of Hibiscus is a native to the United States. It is commonly found in swampy marsh areas, normally in Georgia, Florida,

HIBISCUS
Hibiscus coccineus

and Alabama. Its striking beauty has led to its cultivation in other areas, however.

The flower has been combined with Rose hips to make a pleasant tea. When dried, Hibiscus flowers make a lovely addition to many potpourris.

CHEESE
Malvaceae
Malva neglecta

This small, creeping plant grows to a length of about two feet. It has whitish-lavender flowers from one-half to three-quarters of an inch in width; they have five petals that are notched at the tips. The leaves are somewhat round, about one and one-half inches wide, coarsely toothed, and prominently veined. The plant's distinguishing feature is its fruit, which resembles a wheel of cheese or a sliced pie. Cheese is found almost everywhere; it is generally considered a weed to be found in waste places or disturbed areas. Cheese flowers from April to October.

The young leaves and the green fruit, which is somewhat okralike, are used sometimes as a thickener in soups, or are eaten cooked or raw in salads. Large quantities should not be consumed, however, since such a practice has been known to cause digestive disturbances.

Leaves of the Cheese are said to make a good demulcent tea for treatment of coughs, hoarseness, bronchitis, inflammation of the larynx and tonsils, and irritation of the respiratory passages. Additionally, preparations made from the leaves have been used to treat indigestion, inflammation of the mouth, and eczema. A decoction made from the herb has been used to wash wounds and sores, and a poultice of the leaves has been used for the same purpose, as well as to reduce skin irritation.

LOW MALLOW, Dwarf Mallow
Malvaceae
Malva rotundifolia

Low Mallow is the smallest of the Mallows; its prostrate stem grows to a length of six to twenty-four inches. Its leaves are similar to those of the High Mallow: the shape is somewhat roundish, coarsely toothed, and five- to seven lobed. The flowers are a pale lilac color, trumpet-shaped, and appear from May to November. As with High Mallow, Low Mallow is found along roadsides, fence rows, in waste areas, and in fields. It is also sometimes cultivated, often for use as a potherb.

Besides its prostrate stem, which sets it apart from the other Mal-

LOW MALLOW
Malva rotundifolia

lows, the Low Mallow is the only kind that is self-fertilizing: it doesn't rely on insects to do the job. Otherwise, the herb's properties and uses are exactly the same as those of High Mallow.

HIGH MALLOW, Common Mallow, Country Mallow, Blue Mallow
Malvaceae
Malva sylvestris

High Mallow is a hardy plant, reaching a height of from three to four feet. It is commonly found growing in disturbed or waste areas, in fields, and along hedgerows. The leaves are stalked and appear alternately on the stem; they are downy, light green, and are somewhat roundish with five to seven coarse teeth and prominent veins on the underside. The flowers, which appear from May to October, range in color from pink to bright mauve-purple; they have five narrow petals.

The boiled foliage of the High Mallow has been said to provide a wholesome vegetable, and its fruit, or "cheeses," are also edible.

As for its medicinal properties, it is similar to the Marshmallow, only milder. It is generally used only when Marshmallow is not available, for such ailments as coughs and colds, laryngitis, emphysema, and other lung problems. High Mallow has much the same action on

HIGH MALLOW
Malva sylvestris

the body as Marshmallow: it has been used as an astringent, demulcent, and emollient. And externally, a decoction made from the leaves is sometimes used to wash wounds and treat sores.

One interesting note: the flowers of High Mallow were at one time used in a decorative fashion on May Day by English country people. They were said to have been strewn in front of doorways and woven into garlands.

ORCHIDACEAE
The Orchid Family

Mostly perennials with showy, complicated flowers, the members of this family are often grown as ornamental novelties. The flowers are often strikingly irregular and often spurred. The fruit is a capsule, with many minute seeds contained within. There are about 600 genera and around 20,000 species, found mostly in the tropics, where they grow on other plants. The common flavoring extract, Vanilla, is derived from a member of the Orchid family.

PINK LADY'S SLIPPER
Orchidaceae
Cypripedium acaule

It's a sheer delight to come across this showy pink flower when it is in bloom from April to July. Pink Lady's Slipper grows to a height of fifteen inches, and the flower forms atop a leafless stalk of its own. Its delicate, slipperlike lip petal is puffed out like a balloon. The lip

PINK LADY'S SLIPPER
Cypripedium acaule

petal is also veined with red and has a small crevice on its front. The side petals of the flower are greenish brown, and long and narrow. Two dark green leaves grow about eight inches long, are oval-shaped, and have a fuzzy, veined underside. Pink Lady's Slipper is likely to be found growing wild in the forest, particularly pine forests, where there is a thick covering over rocky outcrops.

Like most members of the Orchid family, Pink Lady's Slipper is commonly grown and appreciated for its beauty. The plants are extremely difficult to cultivate, however, and are becoming somewhat rare in their native habitat. The old adage, "Look, but don't touch," best applies here.

PLANTAGINACEAE
The Plantain Family

The family is mostly composed of herbs with prominently veined basal leaves from which a long, thin flower cluster emerges. The tiny flowers are symmetrical. The fruit is usually a nutlike capsule. Botanists agree on three genera, but estimate anywhere from 200 to 270 species in this group. Members of the Plantain family are found all over the world.

COMMON PLANTAIN
Plantaginaceae
Plantago major

Common Plantain is a commonly observed "weed" along roadsides and in some lawns. A long spike of greenish-white, minute flowers emerges from a rosette of veined basal leaves in this species. This herb may reach a height of over two feet, and it will usually begin blooming in early summer, and continue to early fall. It's found everywhere.

The Pennsylvania Dutch and Amish were said to have used the

COMMON PLANTAIN
Plantago major

ground leaves of Common Plantain to make a mush to feed to babies. The leaves are said to make good cooked greens, and the seeds, reputed to be high in vitamin B, when roasted are said to make a nutritious cracker spread. The seeds are also gathered to feed small birds.

Medicinally, the seeds soaked in water are said to provide a gentle laxative, and the chewing of the rootstock is reputed to relieve toothache pain. Common Plantain has also been promoted, in the form of an infusion or decoction of the leaves, as a remedy for bladder problems, ulcers, ringworm, feminine complaints, coughs, hoarseness, bleeding, and insect bites. Additionally, one herbalist says that an American Indian once received a reward from the Assembly of South Carolina for his discovery that Common Plantain was a cure for the bite of the rattlesnake.

A closely related species, English Plantain (*Plantago lanceolata*), grows slightly taller and has narrower basal leaves. It is a favorite food of rabbits and grazing sheep and has many of the same properties of its cousin, Common Plantain.

POLYGONACEAE
The Buckwheat Family

Members of the Buckwheat family will often have a sour juice contained within the simple-shaped leaves. The plants are mostly herbaceous but sometimes appear as shrubs or vines. The leaves are arranged in an alternate pattern on the stem and, many times, they will form sheaths near their bases. The family name, Polygonaceae, comes from a Greek word meaning "many knees," and is a reference to the nodes that often form on the stems of these plants. The flowers are rather small, and they form in spikelike clusters or sometimes in a head. They are usually symmetrical and bisexual. Usually, they have from three to six sepals that tend to look like petals, but there are really no true petals. All flower parts are attached at the base of the ovary. The fruit is said to resemble a lens; it is small and hard and often makes an abundant food source. Found mostly in the temperate climes of the Northern Continent, there are said to be 40 genera and some 800 species.

PROSTRATE KNOTWEED, Knotgrass
Polygonaceae
Polygonum aviculare

Like its close relative, Lady's Thumb, this herb is commonly found all over the world and is generally assumed to be a worthless weed. Prostrate Knotweed has a creeping, knotty stem that may grow up to six feet in length. It has many branches that bear lancelike, alternate leaves which have brownish sheaths at their bases. The pinkish flowers, which appear from June to October, are tiny, and they form in typical clusters. The herb's seeds are a favorite of small birds, and cows and pigs are said to be fond of its foliage.

Prostrate Knotweed has powerful astringent properties, and was, therefore, utilized as a remedy for both internal and external bleeding. It was also said to be good for allaying the symptoms of diarrhea and other more severe intestinal diseases. The herb was guaranteed by some herbalists to kill intestinal worms.

BISTORT
Polygonaceae
Polygonum bistorta

Bistort has large, oval-shaped leaves with heart-shaped bases. The starchy roots were used in stews by the Indians. The rootstalk is black on the outside, but red on the inside. The stem rises erectly and in an unbranched form; it terminates at a height of eighteen inches, where-upon a dense spike of tan flowers blooms in early summer and again in early fall. The fruit is made up of three seeds which are brown and shiny.

Birds, especially songbirds, are attracted to this garden escape. The common name, Bistort, comes from the Latin, *bistorta*, which means "twice twisted." The root is said to contain gallic and tannic acids, and was once used for tanning leather. The starchy S-shaped roots were often used as a famine food by the people of several continents.

The roots are best when roasted, because that is said to give them a nutlike flavor. Deer and bear in the Rocky Mountains are quite fond of Bistort, also. When the root is ground up, it makes a nutritious flour which has been baked into bread.

Bistort has several medicinal properties. Its prime use is as an astringent, where it has been employed in the control of bleeding, bowel problems, and cholera. It was also used to treat hemorrhoids and no-

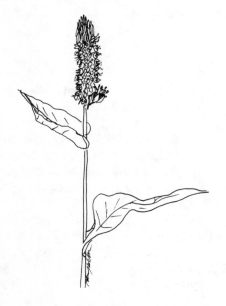

BISTORT
Polygonum bistorta

sebleeds. A gargle made from the herb helped to break up mucous in the trachea. A little of the root and water was used to soothe ulcers and made a good overall tonic, according to old herbalists. The nectar from Bistort's flowers is said to produce good honey. A decoction of the root in wine was claimed to be a preventative of miscarriages and abortions.

Propagation may be facilitated by root division set in the early fall. It will also be successful in the early spring. Bistort is grown in gardens primarily for its showy flowers.

PENNSYLVANIA SMARTWEED
Polygonaceae
Polygonum pensylvanicum

The colorful pink flower spikes of Pennsylvania Smartweed seem to float above a sticky and hairy stalk. The herb grows to a height of three feet. And the bright rosy flower clusters appear from late May to October. This plant's flower is unusual in that it has no petals; but it does have a handful of colorful sepals that are like petals. The leaves are long and narrow. The bases of the leaves form a rounded sheath where the petiole connects to the stem. There are said to be over 30 species of Smartweeds, differentiated by their flowers and their encircling leaf sheaths.

PENNSYLVANIA
SMARTWEED
Polygonum pensylvanicum

Pennsylvania Smartweed is commonly found in woods or waste places. It makes an attractive ornamental plant. You can grow this herb in any fairly moist place in North America. It is found in its native Pennsylvania both in gardens and in the wild.

LADY'S THUMB, Knotweed
Polygonaceae
Polygonum persicaria

This herb draws its common name from the darkened spot in the center of each leaf, thought to resemble the thumbprint of a woman. The plant grows to a height of two and one-half feet and has leaves up to six inches long that are narrow and lance-shaped. There are noticeable bristly hairs at the margins of the tubular sheath that forms at the bases of those leaves. The tiny pinkish or purple flowers form in short clusters at the top of separate stems. They blossom from June to October. Lady's Thumb is found throughout most of the world in waste places and along roadsides.

A common annual, Lady's Thumb has been used medicinally in the past. Its astringent properties led to its use as a treatment for wounds and bruises, but the herb was known to irritate the skin if used to excess. It was also said to have been employed as a remedy for arthritis and other assorted maladies.

LADY'S THUMB
Polygonum persicaria

GARDEN SORREL, Common Sorrel
Polygonaceae
Rumex acetosa

Garden Sorrel is a perennial that grows to a height of three feet. It is found along roads and in damp meadows in much of North America. It was once grown as a potherb. The light green leaves are oblong in shape; those at the base of the plant have stalks, but those at the top have none. Two pointed lobes appear near the base of the stem. The leaves appear opposite on the stem. Branching racemes of reddish -green flowers appear at the top of the plant from early summer on.

The leaves and stem of Garden Sorrel are edible but quite bitter and drawing to the mouth. When cooking them, you must boil twice in separate waters to get rid of the pungent taste. Try them with a dash of lemon juice.

The root has supposed astringent properties, but it has also been utilized in the past as a diuretic and a laxative. A tea made from the leaves and stems of Garden Sorrel was at one time used to help eliminate gravel, but excessive use of the herb has been known to lead to kidney problems and possibly poisoning. The leaves and flowers reportedly ease the pain of mouth ulcers. A tea from this herb was also used to wash the skin.

The plant is a hardy grower and can be started from seeds, if you can wait two years, or from root cuttings if you're not as patient.

SHEEP SORREL
Polygonaceae
Rumex acetosella

An arrowhead-shaped leaf and a tart flavor mark this rather common perennial. The flowers are reddish or greenish, and appear in long, spikelike clusters at the top of the plant. Sheep Sorrel flowers from May to June; the flowers are both male and female. The males are on short, jointed stalks, and a thorough inspection will reveal that the female flowers extend from a deciduous sepal. The seeds follow the fruit and are shiny and brown. The plant gets to be about a foot tall and thrives in acid soil in open areas throughout the North American continent. This hardy member of the Buckwheat family sends out runners from its root, from which it reproduces itself. The runners have a yellow tinge to them, and they grow near the surface of the soil.

SHEEP SORREL
Rumex acetosella

Songbirds flock to this herb because of its abundance of seeds. Although it is considered by many to be a weed and a pest, Sheep Sorrel's flowers can be pleasant to look at when the plants are clustered together. The insect world appreciates this plant and serves as its major pollinator. Deer and rabbits make a tasty lunch of the Sheep Sorrel.

Sheep Sorrel leaves are said to be useable as salad greens, and, in fact, were cultivated at one time for that purpose.

The plant, when ingested, is said to cause a lowering of the body temperature; it also acts as a diuretic. Sheep Sorrel contains oxalic salts, so those with gout or rheumatism may be sensitive to it. It also contains potassium and tartaric acids.

Sheep Sorrel is easy to grow either from seeds or root cuttings taken in the spring or fall. Since it is a hardy plant, it needs little tending. Just keep the weeds away. Keep it in mind as a possible border plant.

MONK'S RHUBARB, Garden Patience
Polygonaceae
Rumex alpinus

Monk's Rhubarb is said to have gotten its name from a reference to its purgative qualities. It's a great plant, growing to a height of four to six feet. It has a thick, hollow rootstalk, and the leaves are quite long and wavy. Reddish or purple flowers appear atop the divided stem.

This herb is a garden escape that now grows along roadsides.

It was the root that was valued medicinally in this plant. It's a slight astringent and was considered good for jaundice and sores. Monk's Rhubarb also has a gentle laxative action.

Seeds or root cuttings will do for propagation of this hardy perennial.

CURLY DOCK
Polygonaceae
Rumex crispus

This member of the Buckwheat family is tall: it grows to a height of four feet. Its leaves are oblong to lance-shaped, in the typical *Rumex* fashion. The veins of the leaves appear rather reddish, and the calyx lobes have toothed margins which are very wavy; hence, the name. The flowers will be either reddish or greenish and will appear in a branching cluster from June to September. The fruit follows thereafter: it is a smooth, brown seed with three sides. Curly Dock will grow in waste places throughout most of the United States and Canada. The young leaves have a lemony, bitter flavor.

Curly Dock leaves can be used in salads or steamed much as you would Spinach. The seeds were ground into meal by the Indians, and a tea of the herb was used to treat skin irritations. The root, when

CURLY DOCK
Rumex crispus

boiled in vinegar, was used in combination with lard or petroleum jelly to make an ointment for the same purpose. The ancients thought of Curly Dock as a bitter tonic and an astringent. It is one of the few plants that supplies iron to the body. The herb must be used when fresh to get the valuable properties from it. The juice of the plant is said to be good to drink when mixed with other vegetable juices.

This member of the *Rumex* genus can be cultivated in a similar fashion to the others. Its height may cause you to consider it as a border plant in your herb garden.

GREAT WATER DOCK
Polygonaceae
Rumex hydrolapathum

As its common name implies, this herb is commonly found along waterways. It is a great plant, and perhaps the largest of all the Docks, reaching a height of over six feet. The leaves are quite large: lance-shaped and narrow, they grow to three feet in length and have a dull green color. The basal leaves are more spear-shaped. Numerous green flowers appear in whorls atop Great Water Dock's many branches. The herb has a blackish root that was used medicinally.

Docks in general are said to be influenced by the sign of Jupiter, thereby making them a blood purifier. The powdered root is said to make a good tooth cleanser in the case of Great Water Dock. The herb is said to have astringent properties, making it somewhat useful in the control of diarrhea. A stomach tonic was made from the sliced root, licorice, water, and a dash of cinnamon. The green leaves were reputed to be an excellent remedy for eye problems, skin eruptions, and freckles.

If you'd like to grow this plant, give it plenty of water, and plenty of room. Use it in the back of your garden, otherwise it may wipe out large areas you may have targeted for less aggressive guests. Great Water Dock is best grown from root cuttings, but it can be started from seeds.

BITTER DOCK, Round-Leaved Dock, Butter Dock
Polygonaceae
Rumex obtusifolius

Another large member of the *Rumex* genus, Bitter Dock is distinguished by its large, heart-shaped leaves. These leaves have reddish veins, calyx lobes, and toothed and wavy margins. The flowers are green and form on whorled spikes that extend from the stem. Bitter Dock flowers are both male and female.

Bitter Dock was introduced to this country and now grows in pastures and roadsides. It is said that cattle will not eat it, however, because of its rough texture. The alternate common name, Butter Dock, comes from the reference to the herb's former use: the leaves were utilized as a wrapping material for butter that was to be taken to market for sale. English common folk also used the leaves to dress scalds and soothe the pain of Nettle stings. One early American doctor, who counted Dock among his medicinal sources, is said to have complained that the herb was so unpleasant that only the jackass, the caterpillar, and man would dare to eat it. But the Indians were very fond of Bitter Dock leaves and stems and considered it among their most reliable food sources. They used it as a potherb along with Sorrel and Lettuce or boiled with Dandelion leaves. And they stuffed their smoking pipes with the dried leaves of Docks.

FRENCH SORREL
Polygonaceae
Rumex scutatus

The leaves of this herb are different from those of Garden Sorrel.

FRENCH SORREL
Rumex scutatus

They are heart-shaped with a notch at their bases. They can also be broken easily and are fleshy and succulent. The flowers of French Sorrel are bluish-gray because of a coating of minute, waxy particles. The flowers are both male and female.

French Sorrel was cultivated in Europe at one time, and the French made use of it in their soups. The new leaves can be cooked and eaten and are said to be quite wholesome. The seeds are ground into meal by the inhabitants of India and by the Eskimos. Many seeds were required when they were boiled, because they tended to lose much of their bulk in the process.

As with most Sorrels, you can cultivate this herb with seeds or root cuttings. Follow the usual procedure.

PORTULACACEAE
The Purslane Family

The Purslane family contains herbs with succulent leaves which are usually alternate or opposite. Where the leaves attach to the stem you'll find a dry, thin, tissue-paper appendage. Flowers are solitary or in clusters in which the central flower of each group is the oldest. The clusters are more or less rounded or flat-topped. There are usually two sepals that remain attached when the flower opens. The four to six petals often fall early. The number of stamens varies: sometimes there is one opposite each petal; sometimes more; sometimes less. There is one pistil with a one-celled ovary. The fruit is a capsule which opens by a circular split like a lid, or it may split open by halves. There are about 20 genera and at least 200 species found mostly in America.

PURSLANE
Portulacaceae
Portulaca oleracea

Frequently a nuisance to gardeners, Purslane is an aggressive succulent with *fleshy, smooth* stems that are thick and reddish in color. Small, pale yellow flowers appear alone or in small clusters. The simple-shaped flowers are about ¼ inch wide; they have five petals, two sepals, and eight or more stamens. The leaves are from ½ to 1 ½ inches long, fleshy, flat, and rounded at the top; they are both alternate and opposite on the stem. Flowers appear from June to November, followed by small, round fruit capsules. The plant can be found as a creeper with branches up to 12 inches long, or may appear as multiple short plants up to six inches tall.

Purslane will grow on cultivated ground or in waste areas all over North America.

The species name, *oleracea*, roughly translated, means "vegetable herb used in cooking." Purslane is high in iron, vitamin C, calcium, and phosphorus. The whole plant is valuable in culinary use. The leaves can be eaten raw, especially in salads, or they can be cooked. Try a nice mustard sauce. The seeds are said to be useful in making breads. The stems, eaten for their moisture content, are said to be excellent when eaten raw. Young shoots can be served as greens or as a substitute for Okra. The plant can be pickled to make a crisp and refreshing snack—a change in pace from cucumber pickles.

Ancient herbalists said that Purslane with a little oil, salt, and vinegar, was cooling on a warm day. When bruised and applied to the forehead, the plant was said to remove heat; when applied to the eyes, was said to remove inflammation; and when applied under the tongue, was said to remove thirst.

Purslane is easily grown in the spring in rich soil. It may be started indoors and transferred in May. By staggering the planting, you can have an ever-ready supply of salad greens. Allow a month and a half for maturity; then harvest by cutting the plants low on the stem so more will appear. Note: Succulents need plenty of water during dry spells. The usefulness of this potherb proves that not all uninvited garden guests are pests.

PURSLANE
Portulaca oleracea

RANUNCULACEAE
The Buttercup Family

Flowers in this family are borne singly, in racemes, or in clusters. They are usually bisexual, symmetrical, and have sepals, petals, and pistils that vary or may be lacking altogether. The flowers usually have many stamens. The leafy herbs bear alternate, on occasion opposite, and usually divided leaves like the fingers of a hand. The vein structure of each leaf is like that of a hand, also. The fruit forms as a berry, a pod, or seed. These herbs are usually found in cooler regions of the northern hemisphere. There are said to be about 35 genera and some 2,000 species.

WOLFSBANE, Monk's Hood, Aconite
Ranunculaceae
Aconitum uncinatum

A hardy perennial bearing several hooded, violet-blue flowers, Wolfsbane blooms from August to October. The flowers are loosely

WOLFSBANE
Aconitum uncinatum

grouped on a nodding, weak stem. They're small and have five petals resembling the shape of a helmet. The leaves are about six inches long and just as wide. They're divided into three- to five-toothed lobes. Wolfsbane grows to a length of four feet, but it usually leans over other supporting plants so that it does not attain that height. The herb is found in low, moist woodlands from Pennsylvania south to Georgia, and in latitudes of a similar range across the United States.

Wolfsbane is listed in pharmacopoeias for its beneficial action in controlling gout, neuralgia, and rheumatism. The entire plant has a nauseating aroma and a bitter, acid taste that is said to make the lips tingle. Its main constituent is an alkaloid, Aconite, which is a deadly poison. Supposedly, arrows tipped with the juice of this herb had the ability to kill wolves, hence the name. Most of the ancient lore concerning this plant refers to the European species, *Aconitum napellus*. It has long been recorded as a quick-acting, deadly poison and ingredient in witches' potions. As indicated, the herb may have beneficial medicinal action when administered in minute doses, but experimentation is not recommended.

Cultivation of the herb is usually facilitated by root divisions, but only a fool would plant this species in a garden of edibles or within the reach of children.

COLUMBINE
Ranunculaceae
Aquilegia canadensis

Columbine has showy red or yellow flowers that appear in early spring, and last throughout the summer. Each flower has petals that curl upward. The overall appearance of the blossom is rather like a group of pigeons stooping for a drink of water. The flowers have five petals, with a hollow red spur that contains a yellow blade. The stamens form a column. The leaves get up to six inches wide and form on a long stalk. They are compound, with many divisions, and are light green in color, bearing three-lobed leaflets. The fruit forms a dry, pointed pod that splits open to scatter the seeds. The plant reaches a height of two feet and can be found on hillsides from Canada south to Georgia and west.

The common name of this herb is derived from the Latin *columba,* meaning "pigeon." The Spanish used to nibble on the root when fasting to quell hunger. The American Indians found the seeds of Columbine to induce labor in pregnant women. They also boiled the roots for a

COLUMBINE
Aquilegia canadensis

tea to stop diarrhea and dizziness and to soothe bronchial and stomach problems. The roots were mashed and rubbed on aching joints.

Plant the seeds of Columbine in the spring of the year in a rich soil with plenty of sun. It's an old-fashioned garden plant, useful for window boxes and flower pots, but it will attract many long-tongued insects seeking its nectar.

GOLDEN SEAL
Ranunculaceae
Hydrastis canadensis

A companion to the mighty Ginseng, Golden Seal is said by many to have just as great, if not greater, medicinal powers. The herb does not command as high a price on the open market as Ginseng, however. One large, wrinkled leaf on a hairy stalk rises from a yellow underground root in this species. With this large leaf appears a single, whitish-green flower in the spring. Underneath the single leaf and flower appear two leaves from the stem, and five prominently toothed and veined lobes make up both the large main leaf, and the separate smaller leaves. Golden Seal, like Ginseng, is found in shady, rich woods. It is common from southern Canada to Georgia, and west in more mountainous regions.

Golden Seal makes a yellow dye and its past medicinal uses are

GOLDEN SEAL
Hydrastis canadensis

numerous. As Ginseng is affiliated with problems of the male, so is Golden Seal associated with the female. The early colonists used it as an overall tonic, and an insect repellent, because of the rank odor of the root. The herb was utilized to stop vomiting and stomach sickness, and it was thought to aid in the treatment of inflamed intestines and hemorrhoids.

To cultivate Golden Seal, you will need a rich, loamy soil with a lot of shade. The seeds require an extremely long germination period and are generally not reliable for propagation. A better way to start the plant is through root divisions: make sure there is a "bud" on each root division. Plant each bud in loose, rich soil and mulch. New shoots should arise in the spring. Set each planting in rows about a foot apart, and allow six inches between plants. Golden Seal is hard to find in the wild, because many people have plundered it without replacing it.

ROSACEAE
The Rose Family

Prickly stems dominate this herbaceous family, but its members include trees and shrubs as well. The flowers are symmetrical and are shaped like a saucer or cup. There are five petals and sepals to the flower generally, but in some cases there are no petals. There is one pistil in some cases, but many in others. The stamens are free from each other and numerous. The fruit is a group or cluster in some cases, but often it is a single seed encased in a hard outer shell. The leaves are alternate on the stem and can be either compound or simple. The leaves are commonly found with stipules at their bases. Botanists claim to be able to identify about 100 genera and around 3,000 species.

AGRIMONY, Sticklewort
Rosaceae
Agrimonia gryposepala

This somewhat tall perennial is well known for its Turnip-shaped fruit, or burr, which contains many hooked bristles that cling to the

AGRIMONY
Agrimonia gryposepala

clothes or fur of any man or animal that brushes past the plant. Hence, the name Sticklewort. The burrs form from the tiny yellow flowers, which are found on a long, erect spike that shoots up from among the pinnately compound leaves. The flowers first appear in July, and they're present only for a month or two. Agrimony is commonly spotted at the edges of fields and woods or amongst the "weeds" observed growing on powerline rights-of-way. It ranges over most of the United States, and, indeed, most of the world, in the form of various species.

The entire plant is fragrant and aromatic: so much so, that its astringent properties seem to be carried in its aroma, irritating the nasal passages of those who come upon it, and causing fits of sneezing.

Most herbals, from antiquity to the present, refer to the European species of this plant, *Agrimonia eupatoria,* which differs from our variety only in size: it is said to be only half as tall as *A. gryposepala.* Nevertheless, both species have similar medicinal properties. Agrimony is said to make a popular astringent tea, which is used both internally and externally for various ailments. It is said to be good for healing wounds, soothing sore throat and ulcers, tightening loose gums, and eliminating persistent diarrhea. It has been recommended for colds and fever and was once used to expel worms from the system. The herb does produce a yellow dye, also.

Agrimony is easily cultivated in just about any soil, using seeds or root divisions in the usual manner.

COMMON STRAWBERRY
Rosaceae
Fragaria virginiana

One of the greatest rewards for a spring hiker is to come upon a Common Strawberry plant, trailing along the ground, and loaded down with its juicy red berries—the kind that melt in your mouth and leave you mad with the ravenous desire for more. It takes two years of growth to form the sweet, seed-laden fruit. From the root with its many runners comes the stem bearing toothed and hairy leaflets. The leaflets of Common Strawberry are found in clusters of three. At the end of a long stalk forms a star-shaped, white flower, nodding from underneath its green umbrella of leaflets. The many stamens and pistils of the flower are yellow, in contrast with the snowy white petals. Common Strawberry may only reach a height of six inches, and it is usually found at woods edges, in fields, and dry open areas. The plant flowers from April to June.

COMMON STRAWBERRY
Fragaria virginiana

WOOD STRAWBERRY
Fragaria vesca

This native herb was a favorite with Colonial gardeners, who did their best to domesticate it. The Strawberry is said to symbolize "perfect excellence." The plant was originally called "Strewberry," because it seemed to be casually scattered about the floor of the forest.

Strawberries are very high in vitamin C, vitamin A, vitamin B, and calcium and iron. It makes a satisfying meal to combine Strawberries, milk, and honey. Also, eating Strawberries will clean your teeth, supposedly, since the citric acid found in them works to dissolve the tartar on the surfaces of the teeth.

The leaves of Common Strawberry, when made into a tea, are said to be good for convalescing children. They can also be made into an aromatic skin cleanser, to clear acne and eczema. Added to other cosmetic elements, they may be useful as a facial pack and are said to relieve sunburn. The leaves were also used to ease urinary tract problems.

Choose a sunny location, with well-drained soil, and use a rooted runner to start your Common Strawberry plant. Be careful to leave plenty of dirt around the root part when you dig it from the ground. Don't expect to see the berries until the second year of growth.

A close relative of Common Strawberry is the Wood Strawberry (*Fragaria vesca*), which has minor differences in its fruit and flower formation. Wood Strawberries are red and heart-shaped, but they bear their tiny seedlike fruits on the surface of the berry, instead of having them imbedded like Common Strawberry. Both Strawberries, however, lend themselves to winemaking. In general, Strawberries found in the wild are found to be of greater medicinal value than domesticated varieties, according to herbalists.

COMMON CINQUEFOIL
Rosaceae
Potentilla simplex

This low creeper attains a height of about six inches on the dry, barren soil it calls home. Common Cinquefoil has distinctive five-petaled flowers that are yellow in color and form atop long stalks that spring from the leaf axils. In this variety, the first flower emanates from the second leaf axil from the bottom of the stem. The flowers are somewhat tiny, and they first appear in early spring and continue to bloom until the early summer. The leaves of Common Cinquefoil are five-parted; the leaflets vary in size, but in this species the largest leaflet attains a length of two and one-half inches. Common Cinquefoil can be found in eastern Canada and most of the United States east of the Mississippi River.

The genus *Potentilla* draws its name from a Latin word for "powerful," or "potent." There are several members of this genus, and

COMMON CINQUEFOIL
Potentilla simplex

most herbalists agree that their presence indicates a somewhat barren soil condition. The reference to power in the genus name of these herbs is said to come from their supposed medicinal virtues. The entire herb is used medicinally, but, for the most part, the leaves are employed in the making of a tea that is said to be mildly astringent. The tea has been employed variously in the treatment of hemorrhoids, sore throat, and mouth and gum conditions.

The roots of some species of *Potentilla* are known to have been eaten in ancient times by civilizations facing famine. For the most part, however, the foliage of these herbs is consumed—not by man, but by grazing animals. Some of the species are too rough even for that, though.

As with most herbs, the leaves of these species should be gathered in the morning just after the dew has evaporated from their surfaces. They must be dried in a sunny or warm place with plenty of ventilation to prevent the formation of condensation.

Another species of this genus is the Canadian Dwarf Cinquefoil (*Potentilla canadensis*), which differs from Common Cinquefoil in that its leaflets are slightly smaller, and its flowers grow from the very bottom leaf axil on the stem. It is found in approximately the same areas as Common Cinquefoil.

A closely related species native to Europe is Five-Finger Grass (*Potentilla reptans*), which is so similar in appearance to Canadian Dwarf Cinquefoil that the vital statistics of the two plants are virtually interchangeable. Five-Finger Grass has a somewhat more colorful history, however. It is reputed to have been a common ingredient in many witches' spells and potions, especially love potions. And, perhaps in a parallel vein, frogs were said to have been quite fond of this herb. Otherwise, the plant's roots and leaves have about the same astringent properties and corresponding medicinal properties as the Cinquefoils.

One other species, Silverweed (*Potentilla anserina*), differs slightly in appearance from the above herbs. This variety is also a creeper, but its leaves are long and pinnately compound, made up of numerous toothed leaflets. Its flowers are somewhat larger, also. Silverweed's range is said to extend to the Arctic Circle, and it is one of the species of *Potentilla* with edible roots.

ROSE
Rosaceae
Rosa spp. (various species)

The sweet scent and beautiful blossom of the Rose has made it a favorite of many just for these properties alone, but it is not only beautiful, but useful. It is also diverse, in that there are more than a hundred known species, each with its own distinct characteristics. But happily, once you get past the prickly thorns, as infamous as the flower is famous, you will find that they all have many of the same uses.

The lovely, fragrant petals of the Rose flower can be crushed to extract Rose oil, used in cooking or to mix in drinks. The petals, steeped in water for at least two weeks, produce Rose water, a useful hair rinse or aromatic tea. Rose water is also used to flavor sherbets and cakes in India, and in the Middle East it is an ingredient in a glaze for fowl. Rose oil is very expensive, but you can make your own. Keep in mind that the oil from Roses in northern climates will be less potent in many cases than that from Roses in southern climes. Collect your Rose petals early in the morning, just after the dew has evaporated. Place them in a glass or ceramic jar and cover them with a fresh, light oil like Sunflower oil. Put the lid on the jar, and set it in a sunny window; give the jar a shake or two every time you pass by. Strain out the petals after a few weeks, and you have your own Rose oil. If it's not strong enough, add some new fresh petals to the same oil and try again. You can use this oil to sprinkle over potpourris and sachets made with Rose petals. The Rose oil, especially from Damask Roses, is used to flavor liqueurs and toothpaste. The fragrant petals of Rose do not lose their scent when dried, nor their lusty color, and so are in great demand for potpourris, sachets, and bouquets. For the strongest scent, pluck the petals just as the Rose is blooming. The petals may be placed in fruit salads and jams, or they may be crystallized for cake decorations. Combining equal amounts of Rose water and glycerine will produce an effective moisturizer and skin softener. Rose water alone makes a sensual bath. Add honey to Rose water for a reportedly good digestive aid. Steep a few petals in vinegar and see what they do for your salad. Finally, the fruit of the Rose, called the "hip," is very high in vitamin C, and can be eaten in different ways. The most popular application of this fruit is in the form of a tea, which is made in the following manner: Gather the Rose hips, trim both ends, and simmer them in three cups of water for each tablespoon of the fruit. Hibiscus flowers are a good accompaniment. Strain out the hips, and sweeten the tea to taste. You can, of course, pop the hips into your mouth right from the bush, much like a Raspberry.

Rose is used medicinally mostly as a flavor masking of other unpleasant preparations. But the flowers are mildly astringent, supposedly, and somewhat of a tonic.

Rose likes a rich, well-drained soil and a lot of sun. You may want to train the plant on a trellis or a border fence. Plant with rooted cuttings spaced at three-foot intervals. You may have a little difficulty getting Rose started, but once it takes hold, it can last for many years. To plant, dig a hole that will easily accommodate the roots without bending them. Make the hole deep enough to bury the whole root clump and part of the stem. Mix compost material with a rich soil, and mound this mixture up at the base of the hole. Position the Rosebush on top of this mound, spreading the roots out over it. Clip back any damaged portions of the roots. Slowly pour in a bucket of water to settle the soil and move it in among the root fibers. Fill the hole with dirt, tamping it down as you go.

One species, Damask Rose (*Rosa damascena*), is quite popular in the United States. It climbs to a height of four feet, and is smothered in white, pink, or red blossoms during the summer. It has large, hook-like thorns and grayish, compound, toothed leaves.

Another species became popular as a living hedge, but is now getting out of control, and borders on being a pest in some areas. Multiflora Rose (*Rosa multiflora*) has white flowers and arching stems that may reach a height of fifteen feet. The thorns are rather flattened, and curved. The leaflets are deeply toothed and have winged leaf appendages. This plant is the state flower of Iowa.

The Virginia Rose (*Rosa virginiana*) bears pink flowers atop a hairy bush. The shrub has scattered, stout, curved thorns, and red hips. It grows to a height of six feet in clearings, thickets, and along shorelines.

ROSE
Rosa

A Rose used to stabilize sandy beaches is the Rugosa Rose (*Rosa rugosa*), sometimes known as the Wrinkled Rose, because the leaflets look like they are wrinkled. The leaves are a dark green, and the flower may be either white or somewhat lavender in color. The stems are very spiny. The fruit is a bright red, with long sepals that remain on the Rose hip.

Still another variety is prized for its fragrant leaves. Sweet-Briar Rose (*Rosa eglanteria*) has smooth, but prickly stems, and small pink flowers.

And finally, the Wild Dog Rose (*Rosa canina*) has pink or white flowers, is taller than the Sweet-Briar, and has a hooklike thorn. It does not have the scented leaves.

RED RASPBERRY
Rosaceae
Rubus strigosus

This common erect shrub is covered with stiff, rigid bristles. The leaves are rather like a feather, having the parts arranged in two rows along a common axis. The leaves appear in groups of three. The Red Raspberry forms a white cup-shaped flower in May, which later forms into a red compressed cluster of tiny, shiny globes. Red Raspberry

RED RASPBERRY
Rubus strigosus

loves to grow in thickets, spreading out over other plants under its four feet in height, or it can be found in abandoned fields, where it appears as a green waterfall. The berries appear from June to August, and they come off easily in the picker's hand when they are ripe. The taste of the berries is sweet, but tart.

A bowl of chilled Red Raspberries with a little milk and honey makes a complete and satisfying meal. You can also crush the berries and add them to vinegar. Use two pounds of berries to one pint of white vinegar. The concoction can be iced and drunk or gargled to allay fever or ease the pain of a sore throat.

A tea made from the leaves is pleasant when drunk by itself, but it also has some medicinal properties. It is said to be a good wash for ulcers, and when chilled and drunk, the tea is said to regulate the bowels. The leaves in combination with the powdered bark of Slippery Elm have been used as a poultice to heal wounds and keep scar tissue to a minimum. But Red Raspberry is most famous for its culinary uses.

Red Raspberry pie is an old favorite that is easy to make. Take a quart of Raspberries, and stir them into a cup of boiling water. Add a cup of sugar, and bring the mixture to a slow boil, stirring frequently. Thicken the mixture with a small amount of cornstarch, and stir constantly. Dump the resulting filling into a pie shell. Place a top shell on the pie, and crimp the edges together. Puncture the top shell to allow the steam to escape while baking. Bake at 350 degrees until the crust is brown. Leftover filling makes a great ice-cream topping. Also, you can add the juice of the crushed berries to brandy or whiskey; letting the mixture set for two to three weeks makes a good cordial.

Red Raspberry prefers a rich, loamy soil in sun or partial shade. It will take three to four years to get berries if you're starting the plants from seeds. The plant is best propagated with root runners. Make sure you allow several feet of clear space around each plant. You may want to train the upcoming long stems on a trellis or fence. In the fall, chop down the old woody stem, but allow a couple of feet for a new start next year. Use protective mulch over the winter.

SALAD BURNET, Common Burnet
Rosaceae
Sanguisorba poterium, Sanguisorba minor

Salad Burnet has a Cucumber flavor and is valued in cool drinks and tossed green salads. A native to the Mediterranean region, the herb will stay green in many southern climates all through the winter, but

SALAD BURNET
Sanguisorba poterium

needs mulch to protect it from the cold. The hardy perennial rises from a flat rosette of leaves into slender, flowering stems lined with attractive toothed leaflets. The flowers are green, but they have red stamens, and they bloom for about three months during the summer.

Introduced from Europe, where it was popular for centuries, Salad Burnet has now been naturalized in the United States. It's an old-timer and was a plant that was used to line walkways so that it would release its wonderful aroma when people walked on it.

The herb makes a nice garnish for wine drinks; the bruised leaves add a subtle flavor if steeped long enough or powdered and added directly. The practice produces a refreshing and cheerful summer drink. Grind a little of the leaves into butter or cream cheese, or chop and mix into sour cream for a cooling vegetable dip. Soak a few sprigs in vinegar, and enjoy the French dressing you can make.

A tea made from the leaves of Salad Burnet in water was said to be a great purifier and general tonic. Taken warm with honey, it was said to ward off contagious diseases and to heal all manner of wounds.

If allowed to go to seed, Salad Burnet will sow itself, or you can begin new plants from seeds planted late in the year or in the spring. Cut back the colorful little rosettes and more leaves will form. If you want only the tenderest of plants, sow this herb every year. Propagation may be facilitated by root divisions, also. Leave plenty of room between the set root parts. The leaves are best harvested in July if you want to dry them. Otherwise, they should be plucked young, when they are tender. With enough sun, the herb makes a good houseplant.

A related species found further north, Canadian Burnet (*Sanguisorba canadensis*) forms a spike of small white flowers, with four showy stamens. The stamens resemble fingers, except that they are quite hairy and downy. The leaves are stalked, with up to fifteen compound toothed leaflets. Canadian Burnet grows to a height of five feet in swamps, wetlands, and bogs, and it flowers in midsummer to mid fall.

RUBIACEAE
The Bedstraw Family

Climbing herbs, shrubs, and trees make up this family. Flowers usually form in branched clusters of symmetrical blossoms, and the leaves are simple, opposite or whorled, and often connected at their bases. The fruits of these plants can be a capsule, berry, or carpel. There are an estimated 500 genera, and perhaps 7,000 species, most of which grow in tropical regions. The widely cultivated Coffee is a member of this family.

SWEET WOODRUFF
Rubiaceae
Asperula odorata

Sweet Woodruff is an attractive creeper with square, shiny stems, whorled leaves, and a bristly fruit. The perennial will attain a height of one foot, and it has the distinctive aroma of new-mown hay when dried. The leaves appear in clusters of six to eight. The flowers appear

SWEET WOODRUFF
Asperula odorata

in early spring, and form white, star-shaped blossoms at the top of the plant. The hairy bristles of the seeds will attach themselves to man's clothing or the fur of passing animals. This herb is cultivated in the United States.

Sweet Woodruff, when dried, is said to repel moths. And in the garden, the plant makes an excellent ground cover. The herb has been added to fruit juices, wine drinks, and teas.

The tea of the herb is said to be somewhat of a stimulant, and is reputed to relieve stomach pain. The herb is thought to dissolve kidney and bladder stones, cure headaches, and sedate those with nervous conditions.

Root divisions set in early fall are your best bet for the propagation of this herb. The seeds require much germination time and are not reliable for starting the plant. Sweet Woodruff must have moist soil and plenty of shade.

YELLOW BEDSTRAW
Rubiaceae
Galium verum

Yellow Bedstraw grows to three feet in height in dry fields and waste places. Its thin, square stem has mossy green leaves that circle the stem in whorls of six or eight. From May to September, the herb

YELLOW BEDSTRAW
Galium verum

forms small clusters of tiny yellow flowers. It is common in the northeastern United States.

The herb is said to have been used to stuff mattresses at one time, but is generally grown as a garden ornamental.

Medicinally, Yellow Bedstraw has been used on wounds and sores to stop bleeding and heal the skin. Additionally, a tea made from the herb is said to be somewhat sedative, and has been used in the treatment of epilepsy.

Root clumps set in the spring are your best bet for propagation. Yellow Bedstraw likes any soil and does well in sun or partial shade. The plant may have to be staked to get it to stand up.

RUTACEAE
The Rue Family

Members of this family are only rarely herbs and are more often shrubs and trees. They have either simple or compound leaves that are leathery and dotted with glands. The flowers are symmetrical, with five petals, and a fleshy disk between the stamens and the ovary. The fruits are berries, capsules on rare occasions, or drupes. Botanists say there are 140 genera and about 1,500 species in this family, of which Orange, Lemon, and Grapefruit are members.

GAS PLANT, False Dittany, Burning Bush
Rutaceae
Dictamnus albus, Dictamnus fraxinella

Its fresh, lemonlike flavor and attractive flowers are overshadowed by another curious trait this herb possesses. Gas Plant produces a vapor that, when ignited, will burst into flame. The reaction usually occurs in hot, dry weather, and does not seem to harm the plant in any way. This long-lived perennial is a bushy, two-foot-high plant, and has in

GAS PLANT
Dictamnus albus

some gardens lived to be fifty years old. In early summer, pinkish or white flower spikes form on stalks that rise a foot above the dark green, leathery leaves. The flowers are not produced until Gas Plant is about four years old, and they occur in great abundance from then on.

In olden times, the herb was used for nervous conditions and fever remedies. Gas Plant tea is said to make a good cordial and stomach remedy. The powdered herb in combination with Peppermint was used in the treatment of epilepsy, and an infusion of the herb was used as a facial wash.

Any good, well-drained soil will suit Gas Plant, especially if there's a lot of sun. The herb is best started from root cuttings, but make sure you place it where you want it to be some thirty years from now, for with a little mulch, and some tender loving care, that's where it will be, still blooming strong. Chances are, that all the leaves will die down over the winter, but take heart, your Gas Plant will reappear the following year.

RUE
Rutaceae
Ruta graveolens

A strong-smelling evergreen, and long a garden favorite, Rue was

RUE
Ruta graveolens

once grown for its medicinal properties, but it is now grown mostly for its bluish foliage and yellow flowers. The herb is a bushy plant that can attain three feet in height and width. It produces leaves with three to five leaflets. From midsummer into fall, small yellow flowers with four or five petals bloom in groups at the end of the stems.

Rue is often called "The Herb of Grace," because of its former use in religious ceremonies. An infusion of Rue leaves and water is said to kill fleas if sprinkled throughout the house. The herb has been utilized to protect against evil spirits in ancient times, and it was thought to be an antidote for various poisons and the bites of insects. In China today, Rue is used to counteract certain poisons and to treat malaria.

You can grow this perennial that stays green most of the year in most places in the continental United States. Rue loves a lot of sun and a soil bordering on alkaline. Like its cousin Lemon, Rue prefers a sandy soil. Seeds may be planted indoors in pots, and the seedlings transplanted later about a foot apart outdoors when the danger of frost is gone. To insure longer flowering, snip off the first blooms that appear. Mulch heavily.

SCROPHULARIACEAE
The Snapdragon or Figwort Family

Members of the Figwort family are mostly herbs. They can be shrubs, however, and, occasionally, trees. Many times the herbs are accompanied by showy flowers and the plant may be partly parasitic. The leaves are green and appear alternate, opposite, or whorled, without stipules. Flowers are usually cymes or racemes. Four or five sepals will usually appear; they will be united. There are also four or five petals that form a corolla with upper and lower lips. Look for four stamens, but there are sometimes two or five stamens. Often, the fifth stamen will be sterile and different from the rest. All flower parts are attached at the base of the ovary. The fruit is usually a two-chambered capsule or sometimes a berry. Many species are found in the western United States. There are 220 genera and 3,000 species throughout the world.

FOXGLOVE, Fairy Thimbles, Lion's Mouth
Scrophulariaceae
Digitalis purpurea

Well known as a heart stimulant, Foxglove, in its second year of growth, bears majestic bell shaped flowers along a five-foot tall spike. The thimblelike flowers can range in color from a delicate white or yellow to a rusty purple shade. The flowers have five lobes, and are beset with black spots, a hairy mouth, and many brownish seed capsules that grow in a pyramid shape. The leaves are four to twelve inches long, two to six inches wide, ovate, and hairy on both sides. The heart stimulant, digitalis, is extracted from the second-year leaves of the plant, just before it comes into flower.

Digitalis is potentially poisonous and should definitely not be trifled with. It should be used only under the supervision of an experienced physician. It is narcotic, sedative, diuretic, and antispasmodic. It was used by physicians in the past to reduce the force of blood circulation. The drug is prepared from the leaves of the Foxglove plant; when good, the leaves are a dull green color, have a feeble, narcotic odor, and possess a bitter, unpleasant taste. The Foxglove flower is a favorite of honeybees, and in fact, the plant's propagation depends almost entirely upon the normal activities of that insect. Other animals, however, will instinctively stay away from the poisonous leaves.

FOXGLOVE
Digitalis purpurea

   Strangely enough, old herbalists knew only that the expressed juice of the leaves of Foxglove was good for external application to various skin conditions. They recommended the bruised leaves, or an ointment made from them, for treatment of wounds, sores, and ulcers.

   The Foxglove plant is cultivated commercially for medicinal use today. Interestingly, one study of the herb showed that plants with purple flowers yielded a higher concentration of purified digitoxin than plants with white flowers or other colors. The study also confirmed a belief that the upper leaves of the Foxglove are much more potent than the lower leaves.

   For those not interested in commercial production of the Foxglove, however, it can be grown successfully simply for the sake of its striking appearance. The plant prefers rich, well-drained soil, and, while it will survive in shady areas, it grows best in full sunlight. Seeds may be sown in early summer, and when seedlings are about one to two inches high, they should be transplanted for eighteen-inch spacing. The young plants should be protected from freezing with a thick mulch over the winter.

EYEBRIGHT
Scrophulariaceae
Euphrasia americana

   Herbalists agree that Eyebright is "an elegant little plant," and its

EYEBRIGHT
Euphrasia americana

genus name, *Euphrasia,* is derived from the Greek *Euphrosyne,* referring to one of the three graces who was reknowned for her cheerfulness.
She is said to have endowed the herb with its valuable properties.
Reaching a maximum height of fifteen inches, it is generally rather
small in stature. The plant has a hairy stem, and it develops clusters
of white and lavender flowers that form on short spikes. The flowers
are from one-third to one-half inch long; the lower lip of the flower is
distinguished by its three notched lobes. The leaves range in size from
one-quarter to three-quarters inch in length, and they are coarsely
toothed. The fruit is an oblong pod filled with many seeds.

   The flowers, which sometimes have yellow and violet stripes, were
thought to look like bloodshot eyes. Indeed, the entire plant was used
extensively to treat diseases of the eye. It was thought to be particularly
beneficial in the treatment of catarrhal opthalmia, but it was also used
for coughs and other catarrhal infections. When combined with Gol-
denseal, Eyebright was said to act directly on the mucous lining of the
eyes. Often, in the late nineteenth century, the plant was simply hung
in rooms to aid weak eyes. The poet Spenser penned:
   "Yet euphrasic may not be left unsung,
   that gives dim eyes to wander leagues around."
The leaves have a slight aroma and a bitter taste. Eyebright tea, said
to remedy hay fever, can be brewed from fresh or dried plant parts.
Because of its supposed tonic properties, Eyebright tea was thought

to be a refreshing and soothing cleanser when used on a regular basis. More than a hundred years ago, tired or inflamed eyes were bathed with an infusion of milk or water and Eyebright. It was said that a poultice of the herb would act on nasal congestion, earache, and hoarseness; it was recommended to ease the discomfort of the common cold.

One recipe for Eyebright tea called for a teaspoon of the leaves to be steeped in a cup of boiling water for about half an hour. The tea was to be drunk cool, but you were to limit yourself to two cups a day. And Eyebright boiled in wine was said to be a great tonic to sip while tucked within the cozy warmth of your bed. An eyewash made from a tablespoon of Eyebright in a cup of boiling water was thought to strengthen the eyes and keep potential ailments to a minimum. The plant was to be gathered in July and August when the flowers were in bloom.

Growing Eyebright can be difficult initially, but the plant is fairly resilient once it is established. The garden must provide Eyebright with the roots of nearby plants in order to insure the survival of this symbiotic herb. Grass roots will do fine. And if allowed to seed itself, Eyebright should return to you year after year, as long as it's in a sunny location.

YELLOW TOADFLAX, Butter & Eggs, Fluellin, Pattens & Clogs, Flaxweed, Ramstead, Snapdragon, Rabbits
Scrophulariaceae
Linaria vulgaris

Walking through the fields from May to October, you will be delighted to find these yellow spurred flowers. The lovely blossoms form in a steeplelike cluster atop a leafy stem. The two-lipped flowers are about an inch long, with five united sepals and petals. The upper lip has two lobes; the enticing lower lip has three lobes with orange ridges and a prominent spur at the base. There is one pistil, four stamens, and a green style. The leaves range from one to two and one-half inches long, and are gray-green in color. You will note two different types. The upper grasslike leaves appear alternately. The lower ones are opposite or whorled. Yellow Toadflax will grow from one to three feet tall, depending on how good the soil is. You'll find it growing wild not only in dry fields, but also along roadsides and in waste places.

The name Toadflax came about because *Linaria vulgaris* was said to be a shelter for toads. Supposedly, there was thought to be a resemblance between the mouth of the flower and the wide mouth of a toad. The herb looks like the Flax plant in the early summer. The genus name is Latin for flax. The name "Butter & Eggs" comes from the plant's bright yellow flowers. There is no culinary use for this plant; in fact, animals won't even touch it. The plant is often strewn in barn stalls to rid them of unwanted pests and insects.

Once reputed to have many medicinal applications, Yellow Toadflax is now rarely used for healing. The whole plant was gathered just when coming into flower and was then used either fresh or dried. There is a wonderful, sweet odor when the plant is fresh which disappears when the plant is dried. Toadflax contains an acrid oil which is said to be poisonous. Besides the yellow coloring, the plant contains two glucosides, tannic and citric acids, mucilage and sugar. In the past the plant has been used as an astringent, a liver treatment, a body cleanser with purgative and diuretic properties, and a jaundice and skin disease cure. The bitter-tasting infusion was used to eliminate excessive body fluids. The essential oils in the herb cause the acrid taste. Seventeenth century herbalists said that bathing in a decoction of Toadflax would take away yellowness of the skin. And the fresh plant was often bruised and applied as a poultice for hemorrhoids. In addition, the same process was used occasionally to make a similar ointment for skin problems. The whole herb was chopped into pieces and boiled in lard until crisp. After straining, a green ointment was formed that was said to be good for sores and ulcers. Toadflax juice and water was considered a good

YELLOW TOADFLAX
Linaria vulgaris

remedy for inflammation of the eyes. In the Middle Ages, the juice in the form of a decoction, combined with alum, was applied to linens as a starching agent. Blueing was not discovered at that time; so, as a result, the Toadflax juice tended to tint laundered goods yellow.

Toadflax is usually gathered in the wild, but it can be cultivated in two ways. First, by separating the roots and replanting them in the fall, you should get a healthy crop the following spring, provided you keep them well-watered. Second, by seeding the plant in the spring, you can start a crop that will, in turn, seed itself in the fall. The herb multiplies rapidly. Its small, rounded fruit is a dry capsule which opens at the top by several valves. When the breeze sways the stem, the seeds are flung to the ground. A European native, Yellow Toadflax adapts well to dry areas. A close inspection of the colorful flower reveals a lower lip which serves as a guide for honeybees and other nectar-seeking insects. Scientific investigation has shown that the orange coloring on the flower heads will attract Hawk Moths even when the flowers are pressed between glass plates. The moths will leave behind tongue marks on the glass in proximity to the orange mark.

GREAT MULLEIN, Common Mullein, Aaron's Rod, Jacob's Staff, Blanket Leaf, Candlewick, Flannel-Flower, Feltwort, Hedge Taper, Mullein Dock
Scrophulariaceae
Verbascum thapsus

This towering yellow-flowered herb reaches a height of six feet. The wooly stem bursts into a spikelike cluster of five-petaled blossoms that are three-quarters to one inch wide. The blossom's shape is nearly regular, sporting five stamens and one pistil. The bottom leaves are thick and velvety, growing up to a foot long. The upper leaves are smaller and stalkless. The yellow and orange flowers appear from June to September, attracting many bees. The plant is normally found in well-drained alkaline soils and grows wild in fields and along roadsides. It is found most places in the continental United States, with the exception of northern Minnesota and the upper region of North Dakota.

Great Mullein was introduced from Europe; it takes two years to flower. Because of its height, it provides a good contrast to the smaller herbs in your garden. Mullein has a fascinating history. It symbolizes good nature, and the leaves are said to be a very good cigarette substitute. In ancient times, Roman soldiers dipped the rigid stalks in lard and lit them to carry as torches in marches and funerals. The leaves

GREAT MULLEIN
Verbascum thapsus

are still used as wicks to this day. Several centuries ago the Indians insulated their mocassins with the leaves of the Great Mullein, and immigrants to this country did the same with their stockings after observing the practice. The leaves and flowers have been boiled to make a yellow dye.

The erect Mullein plant has a great many medicinal uses. A tea from the leaves is said to make a good remedy for coughs, hoarseness, bronchitis, and whooping cough. Flowers steeped in boiling tea are said to help ease pain and bring on sleep. A stronger version of the tea is said to help remove warts; although some sources say that the fresh flowers, when crushed, will do the same. Five drops of the oil in a teaspoon of cold water, taken several times a day, was at one time thought to prevent bedwetting. A poultice of the leaves was used to ease the pain of sunburn and other skin inflammations. Moreover, boiled leaves steeped in hot vinegar and water were also used to treat skin inflammations. The vapor from boiling water with some Mullein flowers in it, when inhaled, was said to open the respiratory system.

Choose a sunny, but well drained soil to plant your Mullein. The herb likes a winter freeze. It can't survive transplanting, however. Therefore, sow your seeds in the fall, while the ground is still warm. Allow about two feet between each plant. Mullein will tolerate sandy or chalky ground. It takes two years to get a mature plant. In the first year, the leaves spread out at ground level, forming a hairy, gray-green cover that lasts all winter. The second year, the majestic spire climbs into the sky and bursts forth in golden splendor.

COMMON SPEEDWELL, St. Paul's Betony
Scrophulariaceae
Veronica officinalis

Delicate, light blue flowers are delightful in the spring. Therefore, Common Speedwell, with its spikelike clusters of lavender blossoms is widely used as ground cover around the house. The tiny plant has prostrate stems with a matlike growth pattern. Speedwell has hairy-stemmed branches from two to eighteen inches long. The downy flowers are only about one-fifth of an inch in length. They have opposite, elliptic petals. The flowers appear from May to July in open woods and dry fields. You'll find the plant growing from southern Canada to northern Mexico, except in Florida and along the Gulf Coast. The flower stalks grow from the junction of the leaf and the prostrate stem.

The plant is said to be named for St. Veronica, and it symbolizes female fidelity. Roughly translated, the plant's name means "of the shops," a reference to its probable sale as an apothecary with reputed medicinal and cosmetic properties.

A tea is made from the leaves. It has a pleasant taste that is similar to Chinese Green Tea. Speedwell tea is common in Europe, where it

COMMON SPEEDWELL
Veronica officinalis

is called "Thé de l'Europe." It is made by pouring one cup of boiling water upon a handful of crushed leaves. After allowing to steep at least five minutes, the tea is ready to drink. It has been used as folk medicine: uses include treatments for liver, stomach, and intestinal problems. Also, it has had applications as rheumatism, gout, bronchitis, cystitis, and numerous other ailment cures. Fresh juice, consumed in large amounts, was said to improve conditions of gout and various skin diseases. Additionally, the plant was used for pectoral or nephritic conditions. There's little reason to wonder why Speedwell has enjoyed a reputation as a healer of all illnesses. It contains a bitter principle, tanin, essential oil, and saponin.

The flowering herb was used to make an infusion which was taken in small quantities with milk throughout the day to rid one of the above-mentioned problems.

Plants may be started from seeds, but to lessen the growing time, separate the roots in the spring, or use stem cuttings spaced about one foot apart in dry, well-drained soil. Plenty of sunlight is helpful, but the plant has been known to grow in partial shade as well.

Prolonged flowering and more growth is achieved by trimming the blossoms as they fade on the stalk. You would do well to separate the root every four years to make sure the plants do not become too densely populated.

SOLANACEAE
The Nightshade Family

The herbs of this family are commonly identified by their trumpet-shaped or star-shaped flowers. The flowers are found with five petals, five sepals, and usually five stamens. All parts of the flower are attached at the base of the ovary. The leaves of the plants in this family are usually simple, and they appear alternately on the stem. The fruits, which are sometimes edible, are usually berries or sometimes two-chambered capsules. The most striking bit of information about the Nightshade family of plants is that its members are often the basis for a number of deadly poisons. It is said that there are about 85 genera and approximately 2,300 species in the Nightshade family, and the majority of those plants prefer a warm climate. Some more common members of the family are Bell Pepper, Potato, Tomato, and Eggplant.

BELLADONNA, Devil's Cherries, Deadly Nightshade
Solanaceae
Atropa belladonna

Atropine, the alkaloid constituent of Belladonna, makes it both a deadly poison and a valuable eye treatment. It is said that as little as one-tenth of a grain of Atropine, if ingested orally by an adult male, will cause symptoms of acute poisoning. Yet, almost no eye operation can be completed successfully without some use of this constituent, which has the capability of dilating the pupils of the eyes to allow easier examination.

Belladonna is a perennial plant. It grows to a height of two to four feet, and the purplish-colored stem grows from a thick white rootstock. The leaves are dark green and they appear alternately on the stem. The flowers of Belladonna bloom throughout the summer. They are somewhat bell-shaped, five-petaled, and purplish in color. The fruit is a berry; it is black and similar in appearance to a cherry. The berries are sweet-tasting and, therefore, potentially attractive to children. They are, nonetheless, quite poisonous.

Symptoms of Belladonna poisoning include: complete loss of voice, doubling over of the body, involuntary movement of the arms and

BELLADONNA
Atropa belladonna

hands, and noticeable dilation of the pupils of the eye. The generic name, *Atropa,* is from the Greek, *Atropos,* the name of one of the Fates, who was supposed to have the power to cut the thread of human life. And the common name, Belladonna, meaning "fine lady," is said to have come from the reference to the use of the juice of the plant by Italian ladies, who placed it in their eyes to create the effect of greater brilliancy, which it did by dilating their pupils.

Belladonna is a strong sedative and antispasmodic. It is said to have no effect on the voluntary muscles of the body; instead, it affects the nerve endings of the involuntary muscles, causing eventual paralysis, as well as delirium. Despite its terrifying effects, Belladonna has been used medicinally in the past. It has been used to treat neuralgia, heart palpitations, intestinal colic, whooping cough, pneumonia, and typhoid fever. The leaves were even once thought to be a cure for cancer.

Should one wish to grow this plant, it is said that it does best in a light, chalky soil. Belladonna seems to prefer dry, sunny weather conditions. Seeds are best sown in early spring, and they are quite susceptible to attack by insects. It is best, therefore, to plant in soil that has been baked or fired, in order to destroy any potentially offending creatures. Seeds should be kept moist during germination, which takes from four to six weeks.

CHINESE LANTERN PLANT
Physalis alkekengi

CHINESE LANTERN-PLANT
Solanaceae
Physalis alkekengi

A cultivated variety of Clammy Ground Cherry, this similar plant has white flowers and a bright orange calyx when its fruit is ripe. It is cultivated for its spectacular appearance.

CLAMMY GROUND CHERRY
Solanaceae
Physalis heterophylla

This small plant is found commonly in dry, wooded areas or cleared waste areas. It grows to a height of three feet at most, and is easily recognized by its velvety, sticky stem. The leaves, which appear alternately on the stem, attain a length of approximately four inches, and are somewhat heart-shaped. The flowers, which appear from June through September, are small, five-lobed, and trumpet-shaped. They are usually greenish-yellow in color, with darkened centers. The fruit of Clammy Ground Cherry is a bright orange heart-shaped berry, which looks something like a tomato inside.

CLAMMY GROUND CHERRY
Physalis heterophylla

The fruit is edible when ripe. The leaves of the plant and the unripe fruit are poisonous, especially to grazing animals. Otherwise, the ripe berries can be eaten raw or cooked into preserves, and the longer the preserves are stored, the sweeter they get. It is also possible to make a tomatolike sauce from the berry preserves.

Clammy Ground Cherry is commonly found in the wild in southern Canada, and the eastern and midwestern United States.

COMMON NIGHTSHADE
Solanaceae
Solanum americanum

This herb is a species native to the United States. It is recognized by its tiny white flowers, which are found in drooping clusters, as with Climbing Nightshade. The flowers have five curled petals, five stamens, and a yellowish central core. The leaves are about four inches long under ideal conditions; they have a thin, oval shape and are toothed. As with Deadly Nightshade, or Belladonna, the fruit is a black berry. The Common Nightshade flowers from June through November, and it is commonly found in waste areas and sunny wooded patches.

The leaves and berries of Common Nightshade also contain a poisonous alkaloid. The berries, which are definitely poisonous while green, are said to be edible when ripe. Those with a desire to experiment

COMMON NIGHTSHADE
Solanum americanum

in that regard, however, should be extremely cautious.

Common Nightshade is found in the eastern and midwestern United States.

CLIMBING NIGHTSHADE, Woody Nightshade, Bittersweet Nightshade
Solanaceae
Solanum dulcamara

As its name implies, Climbing Nightshade is, indeed, a climbing plant. It is a perennial, somewhat shrubby, but with fairly long creeping branches that attach to other surrounding plants for support. The leaves are found alternately on the stem, and they grow to a length of about three or four inches. They have a distinctive lobe at their base. Climbing Nightshade flowers from May to September; the flowers are small, five-lobed, and usually blue or purple in color. They are found in loose, drooping clusters, which are always opposite the leaf stalks. As with Belladonna, Climbing Nightshade produces numerous berries, although they are a bright red color as opposed to the black berries of Belladonna. The berries of Climbing Nightshade start out a bright green at first, and gradually change to an orangish and then red color.

One of the common names of this plant, Bittersweet Nightshade,

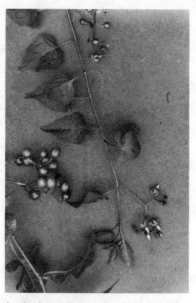

CLIMBING NIGHTSHADE
Solanum dulcamera

is derived from the fact that the root and stem of the herb have first a bitter, then a sweet taste when chewed. The herb has had several medicinal uses in the past, and it has also been used to ward off the "evil eye" from farm animals. A garland made from the branches of Climbing Nightshade, when hung around the neck of the affected beast, was said to be effective to that end.

Climbing Nightshade contains an alkaloid which acts upon the central nervous system. It is mildly narcotic, but if ingested in large enough quantity, it will cause delirium, convulsions, and eventual death.

Preparations made from the herb have been used to treat obstinate skin diseases, as well as chronic respiratory problems. The berries of the plant are known to be poisonous, especially to children.

Climbing Nightshade is commonly found in thickets and at the edges of clearings. It is easily recognized by its bright purple flowers in the summer and by its equally bright red berries in the fall.

TROPAEOLACEAE
The Nasturtium Family

These climbing herbs can be either annuals or perennials. They usually contain pungent juice and round leaves with no stipules. The leaves are attached to their stems at the bottom of the leaf, instead of the margin of the leaf. A single flower usually forms atop a long stalk; the flower is showy, with five petals generally. The upper sepal is a distinctive, long, straight spur. The fruit is a one-seeded carpel that is nutlike. There is only one genus in this family, *Tropaeolum*, which is commonly found around the Gulf of Mexico and south.

NASTURTIUM
Tropaeolaceae
Tropaeolum majus

The round leaves of this climbing annual are set off by a brilliantly orange, red, and sometimes yellow flower. Not only is this plant attractive, but it makes a splashy addition to a salad or sandwich. Nasturtium will reach eighteen inches in height and can be recognized by

NASTURTIUM
Tropaeolum majus

its large, round leaves with their shiny surfaces. The leaves have an unusual formation, in that the stems connect to the middle of the underside of the leaf. The flowers have a little spur on the back, and will bloom all summer until they're killed by frost. Three seeds form in a pod that develops after the flower fades away.

These decorative leaves and flowers are valued as an additive to fresh salads and sandwiches, much as Watercress. They have a peppery flavor like Watercress. The chopped leaves and flowers will spice up ordinary cream cheese. The seeds of Nasturtium, when picked in the early stages of development while they are still green, can be pickled to make an excellent garnish, or they can be added to sauces or relish.

Nasturtium will grow in almost any soil, and some species make great hanging baskets or window-box plants. Others make an excellent border for your garden or fences. The herb requires a lot of sunlight and will respond beautifully to rich soil. It will grow in poor soil, but the plant will go to seed more quickly and have less leaf growth. Each spring, soak the seeds overnight in warm water before placing them in the soil. Plant six inches apart for dwarf varieties and twelve inches apart for climbing varieties. Nasturtium leaves are best harvested before the flower forms. Spray with Garlic water or just plain tap water to dissuade any insects from attacking the plant.

UMBELLIFERAE, or APIACEAE
The Carrot or Parsley Family

The somewhat aromatic members of this family of herbs can usually be distinguished by their fine, fernlike leaf-structure and numerous small flowers grouped in flat-topped clusters. The family contains many valuable food plants, such as Carrots, Parsnips, and Celery; and numerous cooking herbs and spices, such as Anise, Angelica, Chervil, Sweet Cicely, Coriander, Caraway, Cumin, Dill, Fennel, Lovage, etc. The tiny flowers of these herbs are usually symmetrical, with five petals, five sepals, and five stamens, all attached to the top of the ovary. The leaves appear alternately on the stem, and the fruit is composed of two halves, each containing one seed. When last counted, there were said to be about 300 genera and approximately 3,000 species, many of which are native to the United States.

DILL
Umbelliferae
Anethum graveolens

Dill is a hardy annual plant; its round shiny main stem grows from

DILL
Anethum graveolens

two to four feet in height. It resembles Fennel, but can be differentiated by its one main stem, instead of the many Fennel has. Dill has lightly pinnate, finely cut blue-green leaves. The whole plant is rather featherlike in appearance. The flowers appear in a flat-topped cluster; they are a striking deep yellow. The flower, leaves, and stem are all sharply aromatic; the aroma is pleasantly sweet. Dill is a native European plant which has been naturalized in North America.

Dill was mentioned in the Bible, in Matthew 23:23, where Christ said that Dill was scrupulously tithed. The Hebrews used it as a condiment and a medicinal herb. And in the Middle Ages, it was used as a charm against witchcraft. Its species name, *graveolens,* means "heavy-scented." The common name, Dill, comes from the Saxon, "to lull," reflecting the tranquilizing properties of the herb. An old lullaby reads:

"Lavender's blue, dilly, dilly,
Lavender's green.
When you are King, dilly, dilly,
I shall be Queen."

Dill is widely used in cooking. Like many members of the Umbelliferae family, Dill contains an oil of some aromatic importance. The oil is pale yellow in color and is similar to Caraway oil. The herb is used in flavoring of soups, sauces, relishes, beets, eggs (especially when scrambled), and salads. Dill enhances the flavor of fish, pot roast, and lamb chops. A refreshing Dill butter can be made by introducing crushed Dill seeds into butter. The green herb is at its best when combined with cheeses or tomatoes. Dill more commonly flavors pickles and cabbage in the form of sauerkraut. Dill vinegar is popular and easy to make. Just bruise a few Dill leaves and toss them into a jar of vinegar for a few days with the lid on. You'll be pleasantly surprised at the new taste treat. In France, and other European countries, Dill flavors cakes, cookies, and other confections, a culinary art rarely used in this country.

Dill has been used for various stomach and intestinal complaints, especially for infants. A concoction called Dill Water was once a common domestic remedy for children suffering from flatulence. Old herbalists thought that Dill was useful for strengthening the brain, and they recommended its use for curing hiccoughs.

For best results, Dill seeds should be sown in early spring. They should be planted so as to allow at least eight inches between the plants in every direction. The plant is known to draw heavily on soil nutrients, and it is advisable to keep your Dill patch relatively weed-free.

ANGELICA
Umbelliferae
Angelica archangelica

A large, hollow-stemmed plant, Angelica is a biennial reknowned for its many medicinal and culinary uses. Its leaves are found on large branches that emanate from the main stem. The stems of the leaves themselves are somewhat reddish or purple at their base; the leaves are bright green and are formed of triple leaflets, each of which is finely toothed. The flowers, which blossom in July, are found in a large cluster, and are creamy white to yellow in color. The tiny yellow seeds, or fruits, follow after the flowers bloom, and they are quite romantic, as is the entire plant.

Angelica has been cultivated for its stems, which are a popular candied item. They are used to decorate cakes and other assorted confectionery goods. The seeds and roots were once burned indoors to lend a fragrant aroma to the house. The plant is said to bloom on the day of Michael the Archangel: hence, the name. For this reason it was thought to be a protector from evil spirits and witchcraft, and any part of the plant was said to be effective against spells and enchantment. Folklore holds that the Angelica plant is a cure for every possible malady known to man, and, indeed, another legend has it that Angelica was revealed in a vision by an angel as a cure for the plague.

ANGELICA
Angelica archangelica

Angelica has long been used as a sweetener, and a syrup made from the stems and leaves can be stored for extended periods under refrigeration. The roots and stems can be eaten cooked as a vegetable, although the first-year roots of the plant are your best bet. Angelica seeds, and the oil extracted from them, are used as a flavoring for beverages and as an ingredient in the formulation of Vermouth, as well as certain liqueurs, especially Chartreuse.

The entire plant has medicinal properties similar to Anise. As with Anise, the seeds of Angelica are the most potent part of the plant, medicinally speaking. It has been used as a carminative and expectorant, and it is said to be beneficial in the treatment of nausea and flatulence. And while it was recommended for treatment of urinary tract disorders, herbalists warned that it should not be given to patients with a tendency toward diabetes, since it was supposed to raise the level of sugar in the urine.

Angelica can be cultivated in your herb garden simply for the sake of its striking beauty, if for no other reason. The plant prefers a partially shaded area; it will even grow under the shade of overhanging trees. Seeds are best sown in late summer, in light, rich soil that is relatively damp. As with Anise, the plant is rarely bothered by garden pests. Young seedlings survive transplanting well, and for their first year can be spaced about one and one-half feet apart. In the fall, the plants should be moved to a permanent spacing of about three feet.

CHERVIL
Umbelliferae
Anthriscus cerefolium

Chervil, well-known for its spicy flavor, is a biannual plant that reaches a height of about two feet. It has a round stem with fine grooves and multiple branches which contain the delicate, light green leaves. The leaves are opposite on the stem and are fernlike in appearance. The flowers are typical of the family: flat clusters of small white blossoms that appear from May to July. Chervil is pleasantly fragrant. Its seeds, which are elongated and segmented, ripen in August or September.

Chervil was once thought to have the ability to cleanse the blood; an infusion made from the juice of the leaves has been used to lower blood pressure. The species name, *cerefolium*, means "wax-flowered."

The chopped leaves of Chervil are an excellent flavoring for soups, vegetables, and salads. They also make a tasty garnish for egg and cheese dishes, as well as for pork, veal, and beef.

CHERVIL
Anthriscus cerefolium

The expressed juice of the leaves is said to act as a stimulant, an expectorant, and a diuretic. It has been used as a treatment for skin rashes, wounds, infections, and numerous other internal complaints.

Chervil grows best in partial shade and in well-drained soil. The plants come up fast, so with a little planning it is possible to have a fresh supply of the leaves all year round. Seeds are ideally sown in the late spring or early summer. Seedlings should be kept about six inches apart. It makes a great window box plant.

CARAWAY
Umbelliferae
Carum carvi

The name "Caraway" is from the Arabic, *Karawya,* and, indeed, the seeds of Caraway are still referred to by that name in some parts of the world today. The plant is a biennial, growing from about one and one-half to two feet in height. It has smooth, furrowed stems, feathery leaves, and small white flowers which are followed, after blossoming, by the fruit, which contains two crescent-shaped seeds. Flowers appear from May to June and form a compound umbel.

Caraway is highly regarded for its flavorful seeds, or fruits, and aromatic properties, but it was also used as a medicine at one time. It

CARAWAY
Carum carvi

has not only a pleasant taste, both in the leaf and fruit, but when crushed, it has an uplifting aroma to match. It was once an ingredient in love potions where the licoricelike flavor was in demand. The use of Caraway in cooking has been mentioned in the Old Testament, and, to this day, it may be the most widely used culinary plant in the world. Almost everyone has savored the taste of Caraway seeds on rolls, bread, in flour, biscuits, and cakes. Besides the taste, however, was the common superstition that use of the herb would prevent a loved one from wandering away.

In the fall of the year the seeds and flowers of Caraway make an exciting dinner dish. Roasting pork chops with cabbage, apple slices, and Caraway is said to render the meat tender enough to melt in your mouth. Caraway also goes well with soups and cheeses. Even pigeons are said to be unable to resist the delicate flavor of Caraway, and they will supposedly stay close to home when the herb is fed to them.

Caraway is sometimes used as a breath freshener. A "pinch" of the seeds placed under the tongue is said to be quite effective. The seeds will lose their fresh taste and turn bitter if they are cooked too long. But prepared properly, the plant is said to have a beneficial effect on the appetite and digestion. Caraway is also supposed to bring on menstruation, relieve uterine cramps, promote lactation, and act as a mild expectorant.

The herb can be grown from southern Canada throughout most of the United States, except in areas with excessively wet summers. Caraway prefers a sunny place with well-drained soil. Seeds are best sown in early spring when the soil begins to warm up. Caraway should be grown in rows that are about one to two feet apart. Germination takes about two to three weeks. The plants do not transplant well; they should be planted where you want them to remain, with at least eight inches between each one. Remember, this plant requires two years of growth to blossom. The seeds require lots of sunlight to ripen, and they must be harvested before they split open. Seeds should then be dried in a warm place to provide for next year's crop. If you're planning to cook them, however, it's best to skin them to remove unwanted pests before drying. Caraway adapts readily to a window box or other such container, as long as you provide plenty of sunlight. Mulch should be used over the first winter to protect the plants from heavy frost.

CORIANDER, Chinese Parsley
Umbelliferae
Coriandrum sativum

Coriander grows to a height of three feet under ideal conditions. It has a round, finely grooved stem, and, as in other members of this family, it has multiple branches on which the compound leaves are found. The leaves are pinnate, or slightly lobed, depending on their position on the stem. The tiny, delicate flowers are white to red in color and are found in flat clusters. Flowers appear from June to August. The seeds are tiny and round, with a nasty odor until they ripen and begin to dry.

An aromatic and spice, Coriander has been cultivated from ancient times. Its genus name, *Coriandrum,* is from the Greek *koros,* meaning "bug," an indirect reference to its disagreeable odor. The inhabitants of Peru are said to be fond of the use of Coriander in cooking. In Egypt, the round seeds were utilized as funeral offerings to be placed in tombs. And in China, the Coriander plant has a reputation for bestowing immortality.

Coriander is used as an ingredient in curry powder, and the seeds are sometimes used in the making of bread. It is commonly used by distillers in the preparation of gin.

One of the major uses for Coriander seeds is as a flavoring to mask the unpleasant taste of certain medications. It is an integral part of several tinctures, including those of Senna and Rhubarb. Coriander has multiple effects on the body: it is antispasmodic, carminative, and

CORIANDER
Coriandrum sativum

acts as an appetite stimulant. In the past, Coriander was thought to be an aphrodisiac.

Seeds of Coriander are best sown in early spring in a sunny location. Seeds should be planted about one-half inch deep and approximately six to nine inches apart in a light, rich soil. The seeds germinate slowly, but rest assured that by late summer your supply of Coriander will be ready for harvest.

CUMIN
Umbelliferae
Cuminum cyminum

Biblical references are made to this herb, which is grown for its seed or fruit and is a commonly used spice. The plant is small and spidery-leaved. It grows to about two feet tall and has deep green, grooved leaves, which often turn up on the ends. The flavor of the seed is hot and pungent, reflecting its origin along the upper reaches of the Nile River. The plant resembles Caraway; it tastes somewhat the same, but comes on a bit stronger and less refined. The stem has many branches, and the tiny red or white flowers usually appear in June or July. The seed is yellowish-brown, about one-quarter inch long, and similar to that of Caraway. It is thicker in the middle than the seed of Caraway,

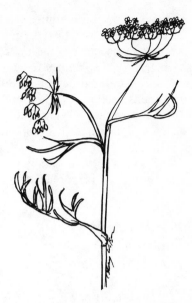

CUMIN
Cuminum cyminum

though, and is lighter in color. Cumin seeds, or fruits, are lighter, hairier, and straighter, with nine ridges which overlap matching oil channels.

Cumin is mentioned in the Old Testament for its use in food and as an additive to flavor wine. Cumin seeds added to 90-proof alcohol was said to be an excellent remedy for congestion. And also in the time of Christ, Cumin was commonly used by the Romans as a currency to pay taxes.

Cumin is one of the ingredients in curry powder. Any stew will perk up in the presence of Cumin. It combines very well with Coriander to flavor vegetable and egg dishes. Texans have been known to use Cumin in their famous chili recipes. And Cumin is widely used in the preparation of both Middle Eastern and Mexican foods. The fruit, when steeped in boiling water, makes an interesting tea. Cumin adds to meat casseroles and will spice up cabbage, kidney beans, lentil soup, or rice. Break those mashed potato doldrums with some crushed Cumin fruit, and try a little in rye breads or cakes.

The fruit can also, when crushed, be added to baking soda for a snappy breath freshener and tooth cleanser. Two to three drops of Cumin oil are said to ease nervous irritability. A teaspoon of powdered Cumin fruit in a cup of boiling water will have the same effect, supposedly, if no oil is available.

Cumin needs plenty of sunlight and a rich soil. Seeds can be started indoors in northern climes, and outdoors in southern regions as soon as the temperature hits the mid-50s at night. When planting indoors, several seeds can be put in one small container to sprout. Then, when transferring the plants outside, take care not to disturb the roots. Cumin makes a dandy border plant. But be sure the soil doesn't dry out during dry spells. To harvest Cumin, wait until the seeds are ripe, but not dry (they'll turn brown), and chop down the plant and hang upside-down in a warm, dry place to complete the drying process. Spread newspaper under the drying plants to catch any fruit that may fall off. When the plant is dry, rub the remaining seeds from the plant with your fingers. The attic is an ideal place to dry your plants.

WILD CARROT, Queen Anne's Lace
Umbelliferae
Daucus carota

This well-known herb, which can be either an annual or biennial plant, has been cultivated from the time of the early Greeks. It is readily observed in the wild, however, in open fields, meadows, and pastures. Its stem, which is covered with sturdy, coarse hairs, grows to a height of one to three feet. The leaves are typical of the Umbelliferae family:

WILD CARROT
Daucus carota

they are fernlike in appearance. The creamy white flowers form a roundish flat-topped cluster. They blossom from May to October.

This forerunner of the modern supermarket carrot that we know and love is remarkably different from its more recent counterpart. Wild Carrot has a whitish, bitter-tasting root with a pervasive odor, in contrast with its reddish-orange, sweet-tasting modern-day cousin. The resultant change is due to crossbreeding begun in sixteenth century Europe.

Wild Carrot is rarely used for human consumption; however, it has widespread use as fodder for farm animals. Wild Carrot has long been valued for its medicinal properties, especially its alleged antiseptic attributes. Wild Carrot soup is said to be a cure for diarrhea; and Wild Carrot in general is beneficial in the treatment of numerous stomach and intestinal disorders. It has been used to treat chronic kidney disease, dropsy, and affections of the bladder. The treatment calls for an infusion of the whole herb to be consumed orally. Its beneficial diuretic action is thought to be caused by its content of potassium salts. And while Wild Carrot probably has just as much, if not more, vitamin A than the supermarket variety, few of us would be willing to ingest its bitter root on a regular basis, but it might help your eyes.

For optimum yield, special attention must be paid to the type of soil that Wild Carrot is planted in. It should be a sandy, light, warm soil. Seeds are best sown in calm weather in the early spring.

FENNEL
Umbelliferae
Foeniculum vulgare

Fennel is a perennial plant; it grows to a height of six feet and has a smooth, shiny stem with many branches. The leaves are somewhat fine and feathery in appearance. The tiny flowers appear in large, flat clusters, are usually bright yellow, and bloom from July to August. The seeds are tiny, oval-shaped, and have the appearance of Dill seeds with the sweet flavor of Anise.

To the ancient Greeks, Fennel symbolized "force" and "strength." The symbolism carried over to the Romans, apparently: Roman gladiators were known to mix Fennel with their food for its supposed ability to act as a stimulant. And at the close of the contests, successful gladiators were crowned with a garland of Fennel.

The seeds of Fennel have a taste like weak licorice and are an excellent companion to meals with fish. And the leaves are used both

FENNEL
Foeniculum vulgare

as a garnish and as a flavoring for salads, stews, and vegetables.

The leaves and seeds, when boiled in barley water and taken internally, were said to increase the yield of mothers' milk and make it more wholesome. The mixture was also the basis for a concoction called "gripe water," which was used to correct flatulence in infants. Seed extracts have been used as a carminative for mild stomach upset and are thought to have the ability to detoxify the body. Also, a solution made from the powdered seeds is said to make an excellent eyewash. Of Fennel, the poet Longfellow said:

> "Above the lower plant it towers,
> The Fennel with its yellow flowers;
> And in an earlier age than ours
> Was gifted with the wondrous powers
>   Lost vision to restore."

All parts of the Fennel plant are said to depress the appetite and are thus considered good for dieters. In the Middle Ages, the plant was used to stave off hunger during church fasts. And the herb has a couple of other unusual properties: the powdered plant has been used as a flea repellent around stables and dog kennels; and an oil extracted from the seeds goes into the making of nylon, plasticizers, urethane foam, lubricants, pharmaceuticals, and certain foods.

Fennel can be grown anywhere, from seeds sown in the spring. It survives well in a dry, sunny situation, but does even better in rich soil.

LOVAGE
Umbelliferae
Levisticum officinale

A European perennial, Lovage is a rather large plant which will attain a height of from three to seven feet. The stem is somewhat round, and hollow, resembling that of celery. Leaves are dark green, deeply divided, and coarsely toothed. The tiny yellow-green flowers bloom in midsummer; they form in flat-topped clusters typical of the family. The seeds that follow are dark brown and oval-shaped. The entire plant is strongly aromatic.

Lovage was once used as an additive to bath water; it was said to be a good cleanser and deodorant. And in colonial times, the root of the plant was often candied for eating.

The leaves have been used as a salad green, and they have been a well-known flavoring for soups, sauces, and casseroles. The stalks and leaves will make an acceptable cooked vegetable, and they can also be candied, like Angelica, for various confectionery purposes.

LOVAGE
Levisticum officinale

Lovage, like other members of the Umbelliferae family, is said to make a good remedy for various digestive difficulties. It is also said to be a strong diuretic: so strong, in fact, that old herbalists warned that it was capable of causing damage to the kidneys.

Lovage can be grown just about anywhere in the United States, as long as there is at least some cold weather in the season: an ingredient necessary for its growth cycle. The plant likes a rich soil, but it will thrive in either bright sun or partial shade. It can be started from root cuttings in the spring, or from seeds sown in late summer or early fall. Whatever the case, the plants need from two to three feet of space between each other.

SWEET CICELY
Umbelliferae
Osmorhiza claytoni

Sweet Cicely is a perennial plant found primarily in heavily wooded areas. It grows to a height of three feet under ideal conditions. The fernlike leaves appear alternately on the stem; they're composed of toothed leaflets ranging in size from two-thirds to three and one-half inches in length. The tiny, five-petaled flowers are white; they form flat-topped clusters. The flowers bloom in early summer.

SWEET CICELY
Osmorhiza claytoni

Sweet Cicely, *Osmorhiza claytoni,* is not to be confused with European Sweet Cicely, *Myrrhis odorata.* The American version's leaves and seeds have a taste like that of licorice, and the roots have a strong licorice odor when bruised. Sweet Cicely is said to impart an excellent flavor to fish.

Sweet Cicely is said to act as a carminative, expectorant, appetite stimulant, and stomach remedy.

ANISE
Umbelliferae
Pimpinella anisum

Anise is a small plant of about eighteen inches in height, with clusters of tiny white flowers and bright green featherlike leaves. It is an annual plant.

Its genus name, *Pimpinella,* is from *dipinella,* meaning "twice-pinnate," a reference to the dainty leaves. Its seeds are its most valuable part. In Biblical times, Anise, along with other spices, was used to pay taxes. The Romans used Anise as a spice, especially in a spiced cake called "Mustacae," which, as well as being somewhat aromatic, was thought to dispel indigestion. Mustacae was usually served at the conclusion of a large feast and particularly after wedding ceremonies:

ANISE
Pimpinella anisum

it is thought to be the forefunner of the modern wedding cake. Gardeners have noted that aphids and other pests don't seem to bother Anise, or any plants near it. And it is said to be an excellent bait for mousetraps, as well as being fatal to pigeons.

The crushed seeds of the Anise are used as a flavoring for soups, cookies, and cakes. They are widely used in the preparation of cordial liqueurs, especially Anisette. The herb itself, when gathered before the seed is produced, is sometimes chopped and used as an addition to cream sauces, salads, and cooking shellfish.

Anise is said to promote digestion, improve appetite, and alleviate cramps and nausea. It is supposed to be excellent in relieving flatulence, especially in infants. Like Fennel, it purportedly promotes the production of milk in nursing mothers; and it has been known to promote the onset of menstruation when taken orally. The herb has probably been more widely used as an expectorant. The seeds have been made into lozenges and are also smoked to promote expectoration. The liqueur, Anisette, mentioned above, is said to be an excellent remedy for bronchial afflictions and asthma attacks.

The seeds of Anise are best sown in the spring; the plant prefers a sunny location, usually in somewhat dry soil. Anise flowers in July, and the seeds ripen in the autumn.

PARSLEY, French Curly Parsley
Umbelliferae
Petroselinum crispum

This quite popular garnish source is a plant that grows to a height of two feet. Parsley is a biennial plant, and it usually attains its maximum height only in the second year. The stem is dark green, deeply grooved, and angular. The leaves, similarly, are dark green and shiny, and they are deeply divided or coarsely toothed. They are extremely curled and are thus quite commonly used raw as a garnish. Flower stalks appear in the second year, and the usually white flowers bloom in late summer. The seeds are brown.

Parsley is a good source of vitamin C. The chopped leaves can be added to just about any dish, especially salads, sauces, vegetable dishes, and fish.

Parsley has been recommended in the past for the treatment of various ailments, mostly those related either directly or indirectly to the kidneys. As with Lovage, herbalists have warned of the potential for damage to those organs through overuse.

PARSLEY
Petroselinum crispum

Parsley seeds are best sown in the spring in rich, well-drained soil. They can take up to six weeks to germinate, though. An old trick to speed the process along is to soak the seeds overnight in water right before planting. Soil should be kept moist, and the new plants should be thinned as soon as they touch. The leaves are ready for use when the plants are about six inches high. Leaves can be preserved by refrigerating, freezing, or drying. Parsley can easily be grown indoors, also.

Chapter 4

Gardening

Growing Your Own Herbs

There is nothing quite as satisfying as a walk through your own herb garden, with its lush green and trembling leaves under the morning dew. Not only do you remember all the effort and planning that went into the first stages, but you can't help but be envigorated by the heavy fragrances of the plants you have chosen for your own. The variegated greens and grays, along with that special fragrance, can easily transform any backyard or balcony into a private retreat worthy of a person of noble rank anywhere. For many people, the joys of the garden go far beyond a good day's work in the warm soil. It is knowing that you can sleep soundly because you have accomplished something fruitful and good, something that responds a hundredfold with its own incredible reward.

Perhaps the real satisfaction of tending your own garden is the very fact that you did it yourself. You have taken the responsibility for one phase of your life. For many people, that personal responsibility has been superseded by a work structure that is all order and control. Rigidly confined to carrying out the orders of a nameless person or machine that one may never meet, some acquiesce. But deep in the hearts of many of those same people beats the desire to take control of his or her own fate in some way. The herb garden, then, fulfills that need, and, more so, will grow and mature through the years. This is now a much sought after phenomenon in these days of pre-made lunches that are ready and waiting at the drop of a name and gone five minutes later. The herb garden not only gives you the pleasure and sensuality of the day, but each plant brings with it a lore and tradition from the past, offering it to you to carry on in your lifetime.

Another big selling point of "growing your own" is that the gardener has total control over everything that happens within the garden, save acts of God and the general hazards of environmental pollution that may exist. For example, if you believe that pesticides are not to be sprayed on plants grown for human consumption, you have the choice of eliminating them and finding an alternative should bugs or disease invade. Sometimes it may mean only sprinkling the garden with the hose, or it may take a stronger mixture of water and crushed cloves of Garlic sprayed in a fine mist to deter unwanted pests. It is your choice.

Not only are the herbs in your own garden readily available to you, but they are also much less expensive than any "store-bought" or dried variety. When you've had your fill of the garden's earthly delights, you may barter them for other needed items or sell them to nearby restaurants that may feel that freshness is the key to their excellence. Some hearty souls have been so overcome by the power of these lowly and simple plants that they give their lives to the gathering, growing, and sowing of the herbs, travelling across the country much as Johnny Appleseed did before them. Their plants may often be found at country fairs, or hiding amongst the new shoots in specialty herb shops. Community cooperatives are trading and buying more of the fresh and frozen herbs now than ever before.

Starting An Herb Garden

Good Soil

Where does one start this new adventure? As always, begin from the ground up. The basics of a good herb garden begin with drainage. Remember that many of our herbs had their origin along the dry and sandy lands of the Mediterranean region. This means plenty of drainage as well as plenty of sun. For the most part, herbs need a great deal of sunlight, and they form their volatile oils with the help of El Sol. There are exceptions, like Ginseng, Golden Seal, and Lady's Slipper, that crave the protective umbrella of a heavy woodland and nothing else. They need a soil that is rich and loamy.

And, indeed, good soil provides the best foliage growth and overall height. Most plants will do fine in lesser soil, but for top performance, use top (soil) material. With less than the best, the plants may not get as many leaves and may instead go to flower and seed quickly. You don't have to go out and buy a lot of fancy potting soil to get rich, loamy earth, but if you are in a hurry, it may be the only way to go.

Building up poor earth may take a little bit of time, but the results will be well worth it. The best way to go about making good soil is by adding vital nutrients and humus back into the soil. A compost heap is the first big step in that direction. Instead of discarding potato peels and the unused portions of carrots, dump them into a single area. That area should be sectioned off with chicken wire on the sides, and if you are ambitious, with bricks on the bottom. The bricks will serve as a resting place for the kitchen garbage, grass clippings, and garden waste and will prevent the invasion of rats and mice from the bottom (give a mouse an inch and he'll worm his way into anyplace). Of course, a homemade concrete base with treated wooden sides will be somewhat more effective.

Layer your compost occasionally with a 10-10-10 mixture of fertilizer, and dump a shovelful of dirt on to help the decomposition take place at a faster rate. Turn the material over periodically—every month and a half—and in a half year you should have useable compost material. Your organic compost should be dark and somewhat dry and flaky.

Rooftop gardeners may want to consider a small compost container made from a garbage can with a few ventilating holes drilled in the sides to allow gases to escape.

While all this compost is brewing—you can start it during the winter months as well as the summer months—take a good look at the space in which you want to locate your herbs. Is the terrain rocky or flat? Does the sun hit all of it all day or only sections at a time? These become very serious questions as you decide what kind of plants you wish to propagate.

Ideally, the soil in your garden should be somewhere at the midpoint between acid and alkaline. Add limestone if it is acid, or add sulphur or iron sulphate if it is alkaline. If the soil is primarily clay, or otherwise not suited for the propagation of our green, leafy wonders, there are a couple of options possible to improve upon what nature has already wrought. One is to dig out a section of earth where you want your garden to be. Dig down two to three feet and lay a bed of gravel on the bottom, so that water can filter down easily. Then return the original soil, making the top foot a combination of soil and compost. Another option involves building raised boxes or other containers so the water can filter down to the former ground level. This is an accepted practice and a convenient one, as well.

City dwellers may have no choice but to find interesting containers for their own herb gardens. Ceramic vases make good containers, as do old whiskey casks and slabs of wood nailed together to make four sides and a bottom. Water these containers from the top and not from a lower saucer, as some plant lovers do. Watering from the bottom may force the roots to lie in water for an extended period of time and develop rot. Yellow, sickly looking leaves are one sign of too much H_2O. So watch out!

Starting Plants From Seed

So now you have the terrain, the sun, and the drainage all worked out. The next step is getting the new plants started. If you want a large volume of plants, the only way to go is to seed them yourself. For one thing, you can't beat the price. Using seeds to grow your herbs is often a slow process, though, so you'll have to be patient. You must also keep the seeds moist—without drowning them—in the germination period. Keeping the container, or bed, in a dark room during the germination period is often a good idea. If it's not possible or practical, however, a covering, like some newspaper or cut-up grocery bags can be just as effective if laid over the soon-to-sprout seedlings.

As soon as the seedlings are through the soil surface, mist them daily so they do not dry out. By the way, a good rule of thumb to go by when sowing the seeds is to cover the seeds with twice the thickness of earth of the seed diameter. For small seeds, that may mean just sprinkling them on top of the soil and patting them once with your finger. Sunlight helps to germinate a few seeds, and most species should come up nicely for you. Sometime between a few days and about three weeks after the seeds have sprouted, remove your paper cover and replace it with a piece of plastic or glass to keep the tender shoots from drying out. This is perhaps the most crucial time for the young seedlings.

Classification of Herbs

Most herbs are classified in a way that will help you understand their properties. Hardy plants will be able to survive cold winter temperatures and come up again in the spring. A nonhardy, or tender, plant will not survive cold weather and may have to be brought inside during the chilly seasons in northern climes. Annual means the plant must be sprouted every year from seed either by your hand or by the natural seeding process. Biennials need two years to form completely and for the flower to bear seeds. In the first year of growth, the leaves form, and then in the second the plant completes its cycle. To insure a continual bloom of biennials in your herb garden you must sow or set seedlings and cuttings each year, so some are in the first year cycle while others are blooming. Perennials need only be given a start and they will be with you year after year, with only a little help in the compost or mulching end of things, or perhaps a few weeding excursions. You will find plants in nurseries classed either as a hardy or tender plant that takes X number of years to flower. For example, Chamomile is a hardy perennial; Rosemary is a tender perennial. Chamomile must be sown each year, while Rosemary will grow for quite a while but should be brought indoors over the winter. Perennials make good border plants when trimmed into a hedge. They will also give you a starting point each year in your garden that you may fill in with the longer-growing biennials or the yearly annuals.

Fertilizing

Sometimes gardeners want to give their plants the very best but hurt them in the attempt to do so. A fresh load of manure from the barnyard makes a fine addition to the compost heap, but should not be added directly to soil for sprouting or anything else. Each load comes rife with its own colonies of insects and bacteria. Sterilize this material before adding it to the garden. This can be done simply enough by spreading some of it on a cookie sheet and popping it into a 350 degree oven for a half hour. This is not high on the list of aromatic experiences, but it does the job.

Transplanting

Do not transplant your sprouts until they have a secondary set of leaves. The first set will open just as the seedling emerges. To transplant, soak the ground with water and allow it to drain completely. When the second set of leaves is well on its way, and the outside temperature is right (with no danger of frost), gently grasp the seedling by its secondary leaf and scoop out the dirt beneath the root, taking a wide enough swipe so that none of the hairlike root fibers are damaged. Then, using the scoop or spoon for support, carefully drop the plant into its new location in a hole large enough to accept the entire root system without cramping it. Water well and tamp down. Some seeds must experience a cold spell before they get the "go" signal for germination. You can help them along by popping them into the refrigerator for two weeks of dark and cold before setting in a sterile growing medium like vermiculite or sphagnum moss. Angelica, Lovage, Clary Sage, Digitalis, Gas Plant, Hellebore, Sweet Cicely, and Violets can be counted among those who like a chilly winter.

Potted Herbs

Some of the perennials or other plants that you want to bring indoors for the winter may be planted outside in the pot they are started in, provided it is big enough to accommodate the expanding root system. Blend the pot into the garden, and prevent unnecessary and traumatic spills by placing the entire pot into a hole in the ground and banking it to make it secure. The pot can be dug up in the fall, after the plant has been severely cut back of its summer growth, and brought into its winter growing area inside.

If space limits you to window boxes or potted plants only, keep in mind that you should turn the pots every so often so the leaves will grow uniformly all the way around, and not just on the side that continually faces the sun. Potted plants do require a little fertilizer once in a while. About twice a month, instead of adding fresh spring water, use a little liquid fertilizer. Some fertilizers that are now available promote bacteria growth, and actually promote the root growth, by setting up a nutrient system that works in harmony with the soil. Chemical fertilizers tend to have the effect of making the roots lazy. The plant feeds directly from the fertilizer and not from the soil. If the fertilizer is used in large areas out-of-doors, this can promote soil erosion, a bigger problem in our country today, according to government surveys, than it was in the dust bowl days of the 1930s.

Cuttings

A faster way to propagate new herbs is by taking cuttings from already thriving plants. Often the tips of growing plants need to be snipped off to make them grow bushier and to produce more flowers. These snips make perfect cuttings for new plants. Place them into small pots or right into the ground. Use a pair of scissors to get a diagonal, straight cut on the stem; then push a common pencil into the soil where you want the new plant to grow and insert the snip. Press the soil in around the new addition and water it well for a week.

Root Division

Root division is another common way to propagate plants. Early in the spring, slide a shovel or fork underneath the plant you wish to divide. After moistening the soil, and after the excess water drains away, carefully lift out the root. Then get rid of as much soil as you can, as gently as you can, by running your fingers down the root fibers. As you do this, gently pull the roots apart. In the center of the clump you will find the old growth. This may be pulled away, or in some more woody plants, cut away with a knife. Take your new clump to a waiting hole, large enough to accomodate the whole root and plant immediately. The Indians used to put a fish in the bottom of their transplant holes for fertilizer. You don't have to go that far, but a little fertilizer will give the roots an extra burst of energy to get them started. Fill in the soil to its former height, tamp it down, and water. Do this in the late afternoon so the plant has a little time to adjust to its new home before the sun delivers its full force. Fall is also a good time to make root divisions, but only after the plant has gone to seed and is in the resting or dormant period. Tarragon can't be started from seeds unless it is the unruly Russian variety, which is not as pleasant as the French variety; so root division is a must for Tarragon. Chives, Valerian, Boneset, Comfrey, and most plants in the Sunflower family respond well to this type of transplanting.

Yet another propagation method employs the stalks. Slice partway through a stalk of a growing herb. The object here is to let the part of the plant you are bending over and pressing into the soil remain attached to the main stalk. Make sure you cut so that you have a section of the stalk that is long enough to reach the ground or a nearby pot. The bent portion of the stalk is then inserted into a small hole in the soil. Fill in the small hole and place either a rock or a clothespin upon the section to make sure it stays beneath the surface. This is an excellent way to begin new grape vines or other trailing plants. Strawberries produce shoots that form their own roots much in the same manner. These new rooted shoots may be dug up and cut from the runner with a pair of sharp scissors and placed in another location.

Herb Gardens

Now that you know how to get them there, make sure you will have the type of herb garden you want before you begin the process. Nowadays most people conform the garden to the idiosyncracies of their lawn, but in a more regal era, the herb garden was a formal affair. It was often symmetrical, and it formed intertwining hedges that could be viewed from a far off balcony in a remote castle. You can design your own formal garden, though. Just make sure you trim the hedges to a specific height, and add some fragrant plants to add spice to the visual ones. The Boxwood hedge was a favorite for the formal, or knot garden, and it blends in well with the herbs discussed in this text.

Steps in designing a formal boxwood hedge garden.

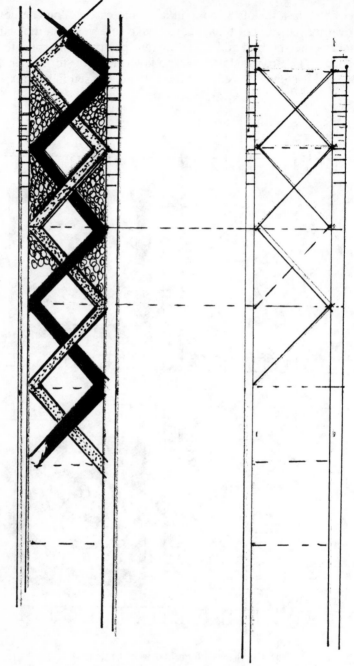

Hedge border for walk or driveway.

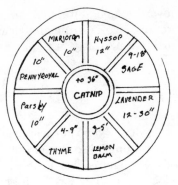

ROSEMARY	PURPLE SAGE	NASTURTIUM	ANISE	ROSE GERANIUM	CALENDULA	ROSEMARY
18"	9-18"	8-12"	2'	4'	4-6"	18"

Plans for a circular herb garden.
Suggestions of herbs to include in a row garden.

Harvesting the Herbs

Harvesting your herbs is one of the joys that accompanies growing your own. The process is satisfying, and the drying plants lend an aroma of freshness to the home, while being pleasant to the eye. The day before you intend to harvest, spray the herbs with a fine mist of water. Your herbs, for the most part, should just be coming into flower. Then, early the next morning, just after the morning dew has dried from the leaves, cut the stalks with a sharp knife or pruning tool. Then you have several choices. The first is to gather handfuls of the herb and tie them together by their stems. Turn the herbs upside down and hang them in an airy place out of the direct sunlight. Direct sunlight will fade the green right out of the leaves. The high reaches of your attic are a suitable place to dry your herbs. Another equally acceptable way to dry the harvest is to strip the leaves from the stems onto a screen. Turn the plant upside down once again, and run your almost-closed hand along the stem, pulling the leaves with you as you go. Spread the leaves out and turn them in a few days to make sure they are thoroughly dried. A faster method of screen drying makes use of the stove. Place the leaves on a cookie sheet and slide it into the oven.

Set the dial for the lowest temperature, and check the leaves often to see how they are progressing. Turn them at least once, and do not remove them until they are perfectly brittle and will crumble in your hands.

Storing Herbs

When the leaves are dry—no matter what method you use—place them in a jar with a tight-fitting lid so the air does not escape and take with it the volatile oils. Sunlight is also a robber of flavor, and a dark glass or large, light-blocking label may be your best prevention against flavor loss and color loss. A good rule of thumb to follow when using your dried harvest says: Take a good deep breath of the herbs when you take them out of the jar for culinary ventures. If the aroma is gone, most likely the oils, and therefore the essence, are gone as well.

Remember that one herb garden will never be exactly the same as another. Enjoy the idiosyncracies of your own, and surely it will offer many years of satisfaction and delight with its leafy and aromatic bounty.

Herbs To Include In Your Garden

"Les Fines Herbes" is a French phrase that refers to the herbs most commonly used in cooking. "Fines" means "highly powdered," in this case; hence, the herbs are finely chopped when added to soups, or other dishes. These herbs are perhaps the ones that you may find most useful in your own home garden for use in the kitchen.

Parsley, high in a number of vitamins and minerals, is a great beginning herb. Soak the seeds in warm water for a few hours before planting, to promote faster sprouting (otherwise, it may take three weeks). Sage, Rosemary, Thyme, Oregano, Basil, Tarragon, and Sweet Marjoram should fill the culinary bill. A few Dill plants for pickle lovers will come in handy. Dill will reach a height of about three feet, so be cognizant of its size in relation to its location among the other herbs of smaller stature.

Mint tea, iced, and with a fresh Mint leaf garnish, is the perfect refresher for a hot summer day, or a pick-me-up, when hot, for a cold blustery evening. So, by all means try your luck with the sun-loving Mint family. Spearmint, Peppermint, and Orange Mint will be excellent starters. You may find that three stalks of each will supply all your needs. The Mints have a tendency to take more ground than you originally intended. To prevent that, a root boundary made of brick, stone, or wood can be constructed before inserting your plants or sowing your seeds. Another way to keep a rein on the Mints is to grow them in pots placed in the ground.

You also might want to consider growing some of the taller species of herbs to act as a border around your garden. The border acts as both a protector from the wind and an attractive focal point.

National Herb Garden

This country's "official" herb garden, The National Herb Garden located at the United States National Arboretum, in Washington, D.C., lists a number of herbs for each of several types of gardens, differentiated by use or historical significance. We've listed some of them here by common names only. Refer to the text for the genus-species name.

Common Culinary Herbs

Dill	Onion
Chervil	Chives
Borage	Angelica
Pot Marigold	Horseradish
Coriander	Tarragon
Sunflower	Samphire
Basil	Woodruff
Marjoram	Bay
Parsley	Lovage
Summer Savory	Horehound
Fenugreek	Lemon Balm
Nasturtium	Spearmint
Oregano	Salad Burnet
Rosemary	Sorrel
Sage	Comfrey
Thyme	Caraway
Fennel	

Common Medicinal Herbs

Aloe Vera
Senna
Licorice
Ginseng
Peppermint
Tansy

Wormwood
Celandine
Golden Seal
Plantain
Mayapple

American Indian Herb Garden

Jack-in-the-Pulpit
Wild Ginger
Dittany
Wintergreen
Pennyroyal
Ginseng
Bloodroot

Indigo
Pipsissewa
Wild Strawberry
Wild Geranium
Bergamot
Mayapple
Goldenrod

Dyeing Herbs

Yarrow
Onion
Dyer's Woodruff
Pot Marigold
Celandine
Lily-of-the-Valley
Sunflower
Elecampane
Broad Dock
American Elderberry
Goldenrod
Tansy

Agrimony
Alkanet
Wild Indigo
Safflower
Lamb's Quarters
Yellow Bedstraw
Indigo
Dyer's Woad
Rue
Bloodroot
Marigold
Mullein

Early American Herb Garden

Yarrow	Anise
Chives	Garlic
Angelica	Chamomile
Tarragon	Borage
Mustard	Pot Marigold
Caraway	Good King Henry
Lily-of-the-Valley	Coriander
Saffron	Gas Plant
Fennel	Yellow Bedstraw
Lovage	Lavender
Woad	Herb Robert
Hyssop	Horehound
Pennyroyal	Peppermint
Bergamot	Sweet Cicely
Catnip	Basil
Sweet Marjoram	Parsley
Rosemary	Oregano
Sorrel	Rue
Sage	Burnet
Winter Savory	Clary Sage
Thyme	Mullein
Nasturtium	Tansy

Chapter 5

Culinary Herbs

Just The Right Amount

The lowly herb has long been used in the finest dishes to bring out the best in a meal. The trick is not to use it to excess. Never try to overwhelm the connoisseur by drowning the flavor of vegetables or meats with too much herb. Often you may find that a half teaspoon of the milder herbs will suit your tastes, and a quarter will be plenty for the stronger herbs. Experiment. Find what delights your own palate and then share that sensation with your friends. You may find that a touch of Garlic is more than enough for your constitution, while another may yearn for a heavier hand. Naturally, each vegetable or meat will agree with certain herbs more than others.

Fresh Is Best

There are a few tips to keep in mind when you set off on your kitchen adventures. Most herbs are at their best when fresh. It is the bruising or chopping that releases essential oils, and it is those oils that give us our smells and flavor, not to mention medicinal properties.

Because the oils tend to dissipate with time, a large supply of the fresh herb is useful. Most herbs can be quick-frozen in a deep-freeze. But shelf storage is also a popular way to have easy access to powdered plants and chopped bits. If you grind your own or prefer the powdered variety, be sure to store your seasonings in air-tight containers that do not let the light in. A label that surrounds a see-through container will help deter some of the light, and thus help to keep your herbs fresh longer. When the scent of your herbs is gone, most of your oils are gone, and it's time to get a new supply.

Recipes

Here are a few select recipes for various herbal delights.

Basic Herb Bouqet

   ¼ cup chopped fresh Parsley
   1 large onion
   ½ tsp Basil
   ½ tsp Chervil
   1 Bay leaf

Chop ingredients together. Mix thoroughly. Fine for sauces, soups and stews.

Bouquet Garni

The following ingredients are placed in a cotton cloth, and added to a pot of cooking food. The pouch can be removed without affecting the texture of a dish you are preparing.

   Hearts of Celery
   Carrot
   Leek
   Bay leaf
   Thyme
   Onion
   Chervil

And how about some new additions to mayonnaise?

Mayonnaise Curry

>   *1 tsp Curry powder*
> *½ tsp ground Turmeric*
>   *1 tsp brandy, sherry, or lemon juice*

Add these to 1 cup mayonnaise, with Chives & Parsley. Mix well. Goes with eggs, chicken, fish, and vegetables.

No-cook Mayonnaise

> *½ tsp Mustard powder*
> *½ dash Paprika*
> *2 egg yolks*
> *2 tbsp lemon juice*
> *2 tbsp white vinegar*
> *1 pt cold vegetable oil*

Mix dry ingredients; blend eggs, then vinegar. Stir or whip mixture while adding oil slowly. When thick, beat in lemon juice. Refrigerate.

Mayonnaise 'n Greens

> *¼ cup Parsley leaves*
> *½ cup Chives*
> *½ cup chopped Green Pepper*
> *½ cup Lamb's Quarters leaves*
> *2 tbsp lemon juice*

Puree all the ingredients in a blender. Add to mayonnaise. Great for cold sandwiches.

   Are you tired of plain, ordinary butter? Try these.

Tarragon Butter

   Mash a teaspoon of crushed, dried, Tarragon leaves into a half cup of butter. Refrigerate for 30 minutes, then roll into balls. Store in glass jar in refrigerator, or freeze. Great for steak or fish fillets.

Chervil Butter

   Cream a few teaspoons of unsalted butter or margarine. Add grated lemon rind and enough chopped Chervil to make the butter green. Roll into little balls when mixture solidifies again in the refrigerator. Great for topping vegetable dishes or fish platters.

Caraway Butter

Blend 4 teaspoons of the seeds of Caraway into a pound of butter. Great for all vegetables. The seeds can also be added to sour cream.

Borage Butter

Mix 2 cups of finely chopped Borage leaves and some Borage flowers with an equal amount of sour cream, and add 2 tablespoons of lemon juice. Salt and pepper to taste, and use on any fish dish.
Well, you get the idea. Here's a general recipe to cover all the bases.

Cold Herb Butter

Served cold, as its name implies, this creation, filled with the herb of your choice, goes great with most hot foods. It's a great way to retain the flavor of herbs throughout the winter months. Simply mix the herb with an appropriate amount of butter (or margarine), add a little lemon juice, and refrigerate. When the mixture returns to a proper consistency, roll it into little balls and refrigerate or freeze them. Experiment!

Of course, no one can survive without the familiar taste of Mustard. Try these variations for a new taste treat!

Basic Mustard

In 20 minutes you can come up with a basic Mustard by crushing some seeds and mixing them into a paste with plain old water. The longer it sits, the more flavorful it becomes.

French Mustard

In a glass jar, combine 2 sliced Onions and 2 cups red wine vinegar, and let stand covered all day. Then strain mixture into another container. Separately, combine 1 teaspoon each of Parsley, Basil, and Oregano with ¼ teaspoon of Cayenne Pepper, 2 teaspoons salt, 1 cup of dry English Mustard, and ½ cup of your strained vingear mixture. Slowly bring entire mix to a boil in a saucepan, adding Mustard last, along with remaining vinegar. Simmer and stir for 5 minutes, then cool.

Stir in 1 tablespoon Olive oil and 2 tablespoons Honey. Mix well, preferably with a blender, and store in covered bottles.

What about sauces?

Mint Sauce

Try this fresh addition to your lamb dishes. Take ¾ cup of your favorite chopped Mint leaves, ¼ cup white wine vinegar, 1 tablespoon sugar, ¼ teaspoon ground Ginger, and 1 teaspoon lemon juice. Combine all of these ingredients in a blender and mix well. Chill the mix overnight, and it's ready.

Mustard Sauce

Make a paste of two mashed egg yolks and 2 teaspoons of dry Mustard combined with 2 tablespoons of Tarragon vinegar. Add 2 teaspoons of the Herb Bouquet we discussed earlier. Melt one cup of butter or margarine and mix well with the Mustard paste. Salt and pepper to taste.

Soups, you say?

Vegetable/Herb Soup

Sauté some mushrooms and onions in a skillet, and add them to a mixture including 1 teaspoon minced Savory, ½ teaspoon Dill seeds, ¼ teaspoon minced Marjoram, 2 ½ cups stewed Tomatoes, 1 teaspoon sugar, and 1 crushed Garlic clove. The herb and Tomato mixture should have been sitting overnight in the fridge. To the whole works add 2 quarts of water and 4 cups of your favorite chopped vegetables. Bring this to a boil, then reduce heat and simmer the soup for several hours. If you like, you can add a cup of long-grain rice for the last 20 minutes you simmer the soup.

Watercress/Borage Soup

Sauté 2 cups minced onions in 3 tablespoons of butter, with a couple tablespoons of flour for a thickener. Don't brown the mixture. Add 2 ½ cups chicken stock and a cup each of Watercress and Borage leaves. Throw in 2 ½ cups of water, and simmer the soup for 30 minutes. Puree the soup through a vegetable mill, and thicken it with 2 egg yolks, and ½ cup of heavy cream. Heat up again to serve.

Want something really unusual?

Deep-Fried Parsley

A surprising delicacy from the East. You'll need a cup of flour, 1 egg white, some warm water, and some vegetable or Caraway oil. Mix the flour, the egg white, and ⅔ cup of the oil with enough warm water to make a creamy, light batter. Dip in the fresh, ice cold Parsley sprigs, and immerse them in hot oil until golden brown. Top with Caraway or Anise seeds. It's great!

Mint Ice Cream

Start with any amount of vanilla ice cream. Add some fresh Mint leaves, from a Mint of your choice, and some crushed ice and lemon juice. With a little experimentation, you'll come upon a "perfect" mix. Toss it all in a blender and go to town. Also, try substituting Rose hips for the Mint, or combine the two.

Mint Slushy

Place a handful of fresh chopped Mint leaves in a pan with 1 ½ cup water and about 8 tablespoons of sugar or honey. Heat until the sugar dissolves, and taste-test to see if the mix is Minty enough for you. Simmer longer for stronger taste. Strain the mix and add the juice of one lemon or orange. Place the new mixture in the freezer until partially frozen. Then add a beaten egg white to the mix, and return it to the freezer until it solidifies somewhat. It's ready! You can also add things like rum or brandy when you add the egg white, or substitute Elder flowers, Geraniums, or Rose petals for the Mint.

Italian Seasoning

This is great with Tomato and pasta dishes. Mix 2 teaspoons of dried Oregano (use the *real* thing), 2 teaspoons of dried Purple (Spicy) Basil, ½ teaspoon of dried Rosemary, ½ teaspoon of ground Sage, ½ teaspoon of dried Savory, and 2 teaspoons of dried, grated lemon rind. Add an equal amount of dried, chopped Onions to the above and bring a homemade pizza to life! Try it on veal dishes also.

Fines Herbes

Last but not least, this is another useful combination delicious with many culinary efforts. Combine ¼ cup of chopped Parsley with an equal amount of Chives. Chop fine. Then add all of the following, or just one ingredient at a time, or try others:

  ½ tsp minced Basil
  ¼ tsp minced Thyme
  ½ tsp minced Chervil
  ¼ tsp minced Tarragon

As you can probably see, the fun of cooking with herbs is in the experimentation process. You can choose what you want to create a new and unique taste sensation. We recommend that the novice herbalist stick to the more common varieties for cooking, however. If you want to experiment with something bizarre, it's a good idea to have an experienced forager advising you, be aware of what you're dealing with, and have the phone number of your family physician or nearest hospital within arm's reach.

Chapter 6

Herb Tea

An invigorating way to start a new morning, or a soothing ritual to end the evening, teas have never been more popular in the history of both the United States and Canada. Part of that popularity is an increasing awareness of the pleasures that teas can bring. There is a new demand that the general population is crying for—herbal tea options. No longer are housewives and top business executives content to brew the conventional East Indian varieties. Experimentation with new taste sensations is in vogue. And there's also the desire to make your own and to know that tea is not some rare or exotic fancy that you must travel great distances at great cost to satisfy.

Most North Americans begin the day with a warm cup of coffee or tea. Caffeine is becoming a more bothersome element than many would like, today, especially when they choose to ingest it on a daily basis. It is an element that is hard on the stomach lining and has an effect on the nervous system. Not only can caffeine give one the "jitters," it can cause a dependence. It is just that sort of chemical dependence that people are tired of. They want to know that they can face the challenges of the day of their own volition, and not at the mercy of some chemical component.

There have always been teas in America. Chinese teas were quite popular until a certain Boston Harbor party made folks think twice about drinking them. The Chinese make two different kinds of tea: green, and black, which are made mainly from the leaves of two plants, *Thea bohea* and *Thea viridis*. Green tea is made from the dry leaves of these herbs; black tea is made from the same plants, but the plants have been subjected to a fermentation process that occurs when the leaves are allowed to dry in the air. It is the fermentation that makes the black color, and, many people say, a highly enriched flavor.

Making Tea

There are three ways to make tea: boiled, whipped, and steeped. Essentially, tea is the combination of boiling hot water with the leaves of various herbs and plants. The leaves yield their volatile oils to produce either flavor or medicinal virtues, or both. Tea making can be honed into a fine art. There are a few basic rules to follow for "perfect" tea, no matter how involved one becomes in the brewing process. First, warm the tea pot with hot water before brewing any tea. Use a ceramic or glass tea pot only. Always use fresh, cool water—preferably nonchlorinated.

Root and bark tea usually has to steep a little longer than teas made from leaves, but try not to allow your tea to oversoak. You can, however, steep your tea in the sun as a sort of "solar tea." Just place the herb's leaves or flowers in a large glass jar in the sun, add cool water, and allow the mixture to steep for a day or two, until the color of the leaves or flowers filters into the water. You can serve the solar tea over ice, with a garnish of mint, or you can take it at its natural temperature, a practice less shocking to the system.

As the Chinese were said to advise, when you're home alone, or when close friends come to visit, then is the time to relax in a comfortable room, with familiar surroundings, sweet memories, and a warm cup of tea.

Teas To Suit All Tastes

We recommend the following herbs, or parts of herbs, for new taste sensations. Simply add about a teaspoon of the dried herb of your choice to boiling water (a cupful), or about three teaspoons of the fresh herb. Allow the mixture to steep until it suits your taste.

Hibiscus flowers	Lemon Balm leaves
Rose hips	Bergamot flowers
Peppermint leaves	Catnip leaves
Licorice root	Dill flowers & leaves
Chamomile flowers	Goldenrod flowers
Elder flowers	Hyssop flowers
Fennel seeds	Wild Thyme leaves
Coriander	Wintergreen leaves
Borage leaves	Lavender flowers
Fenugreek leaves	Pennyroyal leaves
Star Anise leaves	Nasturtium flowers & leaves

Try them all, and see which ones you like, or create your own. Just make sure someone else has tried it first, though, if you're planning to delve into heretofore unknown varieties.

Chapter 7

Herbal Potpourris, Cosmetics, Dyes

Potpourris

A potpourri is a collection of dried leaves, flowers, and spices from herbs. Its inviting aroma can fill a stale room or closet drawer with new and invigorating life. Color, shape, texture, and aroma are the main ingredients of any potpourri. You add what you like, to say what

Potpourris, stored in glass containers which are easily opened, make attractive air fresheners.

Tuck potpourri wrapped in small plastic bags into larger gift packages for a surprising treat.

you want to say, or simply to please yourself. It is the combination of dried flowers and essential oils that make these air fresheners interesting and attractive to many people. Decide what the theme, or main scent, will be, then build upon it with complementary leaves, or try to contrast smells with other fragrances. Another ingredient you will need is a fixative, usually Orris root or common table salt, or gum benzoin. The fixative does two things: mixes the aromas, and holds the aromas there for a long time.

Drying Rose Petals

Rose petals are quite popular for the making of potpourris. Wild Dog, Sweetbriar, and Eglantine all bear flowers that lend their sweet fragrance to the making of dried mixtures. One of the keys to a successful mix is the proper drying of the Rose petals. Keep them away from the sun! The sun may help to form the essential oils that enhance the smell, but once the blossom is plucked from the vine, Old Sol only serves to dry out and evaporate those valuable oils. Pull the petals from the flower and place them atop a sheet of paper on a screen, in an airy dark room, until they are dry and brittle. If you want to speed up the process, place the leaves on a tray in the oven, and set it on the lowest possible temperature for about half an hour. Check the petals at that time, and see if they crumble in your hands when you try to pick them up. If they do, they're done. If they don't, simply let them bake for a while longer. Don't turn up the heat, because that will make the petals lose color and fragrance. With your basic Rose-petal mix complete, you're now free to work with other flower and leaf additives that can be combined in different proportions to suit your nose.

Types of Potpourris

There are two kinds of potpourris: wet, with a little brandy or orange peel; and dry, with just the leaves, and perhaps a dab of oil. The wet potpourri is at its best when stored in an earthen jar or glass jar with a lid on. Small holes in the lid allow the haunting aromas into the air. Dry potpourris are at their best when scattered throughout the house or office in open containers. Naturally, if the mix is located in an area that becomes heated, the aroma will travel faster. That is why you'll often find potpourri mixes atop the mantle or suspended over woodstoves. But the magic of herbal frangrances isn't produced in a moment. The mixes need time to blend their scents. Chamomile and Costmary also make good herbs for generating a lot of fragrance. So are Bay, Bergamot, Lemon Balm, Lavender, Lily-of-the-Valley, Lemon Verbena, and Violet. There are a number of others, and most any plant whose aroma pleases you will do. Experiment to find out which lose their fragrance, and which are enhanced, with drying.

A few different merging scents are okay, but if you have too many they tend to work against each other. Beginners may want to limit themselves to two to four separate fragrances at first, and branch out to more later.

Rose Petal Potpourri

A commonly distributed potpourri can be made easily at home with the Rose petals you gather from the front yard. Here's another way to dry the petals and form your mix. Layer the bottom of a ceramic or glass container with the petals. The layer should be fairly thick, from three to five inches. Then sprinkle on a thin layer of common salt, much as you do when making a casserole. Continue layering the Rose petals and salt until you reach the top of the container. Cover it and place the container in a dark spot for about a week and a half to allow the ingredients to settle. Then, mix up the layers with a spoon and add powdered Orris root for a fixer. Use about a tablespoon per pint of the mix. The Orris root has a Violetlike scent and adds a creamy color to the mix as well. In addition to the fixer, you may want to add other ingredients at this time. Small bits of dried orange peel and one of the following spices should do just fine: Anise, Allspice, Cinnamon, Cloves, Mace, Corriander, Tonka Beans, or a few Vanilla pods. Mix the batch all together and seal again for a month and a half. When the fragrance seems to have reached the desired strength, scoop out some handfuls and place them in little jars around the house for a breath of summer in the cold of winter.

Bay Leaf Potpourri

Bay leaves add a strong aroma and a dusky green color to pot-pourris. Mix this herb with others of a more citrusy nature for a hint of the deep woods or high seas. Try this mix:

3 oz broken Bay leaves
1 oz Lemon Verbena
1 oz Lemon Thyme
3 oz Rose hips
1 oz Safflower petals
1 oz Cornflower petals
1 oz Calendula flowers

Sprinkle all the ingredients with five drops of oil of Pennyroyal, and a half ounce of powdered Orris root. Place the covered earthenware jar, in which you have placed all of the above, in a dark place for about a week, stirring once a day. Then, place the completed mixture in a ceramic bowl at an appropriate location in the house.

Peppermint Potpourri

Peppermint makes a refreshing theme for a potpourri, and it has a similar effect on you after a long day's work. This mix might interest you:

4 oz Peppermint leaves
2 oz French Lavendar
1 oz Marjoram leaves
1 oz Lemon Verbena
1 oz Rose hips
1 oz Star Anize

Set this mix with a half ounce of Orris root, and see if it doesn't add a breath of spring to a stuffy room.

Garden Mix Potpourri

And try this one for a refreshing morning wake-up from the garden:

4 oz Marjoram leaves & stems
4 oz Rosemary leaves
4 oz Lavender flowers
4 oz Rose buds
1 oz Pennyroyal leaves & stems

1 oz Sage leaves & stems
1 oz Cornflower petals
1 oz Bergamot petals

Fix this mix with an ounce of Orris root and six drops of Bergamot oil. Try placing some of this mixture in a cheesecloth bag and hanging it under the bathtub faucet for a tingling bath.

As you see, the combinations and possibilities are limitless. This could become a satisfying hobby for those with an inquisitive nature.

Cosmetics

If beauty is only skin deep, then you obviously want to keep your skin as healthy and glowing as possible. No matter how "beautiful" you are, you'll only feel good if your complexion is as fine as you can make it. Herbal cosmetics might interest you, but when making them, use only pure ingredients. When water is called for, spring water is suggested, to be sure there is no chlorine or fluoride. Fluoride may do wonders to prevent cavities, but it hardly does a comparative job on the face.

Herbal Facial

This recipe for a youthful facial glow comes to us from Denver, Colorado:

Toss a couple of handfuls of Chamomile flowers into a quart of white wine vinegar, and cover. Let it stand for two weeks until the vinegar turns yellow. Then strain the mixture through a tea strainer or two layers of cheesecloth. Pour into a covered jar to store. When you want to use the rinse, add a little to a sink of hot water, and splash it on your face. Your skin will tighten and glow. The vinegar tends to dry the skin, so the mixture may be used more often by those with oily skin problems. The mix gives your skin an acid balance.

Herbal Facial Splashes

Yarrow also makes a healing facial splash when combined with vinegar. But you may try some herbs in combination to find the splash you enjoy the most. Those in the know recommend the following ingredients for facial splashes:

Bee Balm flowers and leaves
dried and powdered root of Blue Flag

    Boneset leaves and flowers
    chopped Borage leaves
    bruised leaves or powdered dried root of Great Burdock
    Calendula flowers for a strong yellow color
    any of the Mints
    Comfrey roots, leaves, and flowers
    crushed root of Elecampane
    Hyssop root

And there are many more. Look for herbs with mildly astringent or emollient properties, and see what you come up with. Additionally, the entire body can benefit from a combination of the above plants and water. Let a mixture of your favorite herb and water stand in the sun for a few days. Strain the mix, and add the herb water to your bath for a pleasant way to relax after a long trying day.

Herbs For the Bath

An easy way to add herbs to your bath is to boil the selected plants in a regular pot on the stove, making sure they're completely covered with water. About eight to twelve minutes should do. Just strain the flowers out, and dump the herb water into your bath. This method will fill the house with a wonderful aroma. You can cut down on the preparation time by simply hanging a bag of herbs under the faucet, as we mentioned before. Try chopping your herbs in a blender and mixing them in a container of baby oil. Use a glass jar with a tight lid. Place the mix in the sun for two weeks, and shake it once in a while in passing. Strain the finished mixture and add it to your bath. This is a wonderful aid for the complexion, and especially beneficial in the drying winter months.

Dyeing

Natural dyes are increasingly becoming an area of interest to many people. While it is true that modern package dyes are simple to use for the most part, are versatile, and have the quality of being long-lasting, there are some who feel that the colors produced by natural dyes are often warmer and more subtle in comparison with those packaged varieties, and that there's a certain comfort in knowing that you have created those colors yourself.

Dyeing a Piece of Wool

The home dyeing methods are not as complicated as you may

think, but they do take time. Here's how it's done with a very common herb, Queen Anne's Lace, or Wild Carrot. Fresh herbs are gathered and placed in a large enameled pot in which they are bruised or chopped and covered with water. The pot sits overnight as the plant's juices combine with the water. Rain water is generally recommended, since tap water is often too harsh and contains chemicals that may affect the color of the dye.

While your dye is brewing, you must prepare your material—wool, in this case. Beginners may wish to start with old-fashioned wool that has been cleaned of its oils. The wool must be soaked with a material that allows the color from the herb to enter the fibers of the fabric and stay there. This material, of which vinegar, alum, and cream of tartar are good examples, is called a mordant. Two ounces of alum or cream of tartar for every two gallons of water can be mixed and heated. Place your wool, in skeins to prevent tangling, into this mixture. The wool should soak for at least ½ hour.

Then begin heating your dye plant mixture slowly. Simmer the mix for about ½ hour, stirring constantly, and then remove it from the heat and strain out the plant material. You can then return the strained dye mixture to the stove.

Meanwhile, remove the wool from the mordant and let it drain in a strainer in the sink, until most of the excess water is gone. Don't let it dry out completely, or it may not "take" the dye evenly. Also, don't squeeze or wring out the wool.

The next step is to insert the wool into the dye mixture and simmer it on low heat until the wool reaches the color you want. In this case, it should be a lovely yellow. If you add a teaspoon of powdered iron, green will be the result. If you have more time, after bringing the

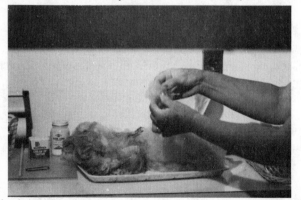

While the dye is brewing, prepare your material—wool in this case—by soaking it with a mordant.

mordant water to a slow boil (160 to 190 degrees F.) for a half hour, let it slowly cool down and sit overnight, and then reheat it the next day. You might also try rinsing the wool when it is drawn from the dye mix and placing it in a pot of vinegar water for a final color set. Repeat the dyeing process if you desire a deeper color.

Keeping Records

Dyeing with herbs can be a bit tricky, and you'll find a lot of factors can influence the eventual result of your work. So it's a good idea to keep meticulous records of what you do and exactly how you do it. Record the type of material you're using, the mordant, the length of time you "cooked" the herb, the length of time the material spent in the mordant, and, of course, the type of herb you're working with. This way you'll have a ball park idea of what to expect the next time, although each batch will probably come out a little different. Eventually, you'll be able to gauge the amount of plant material you'll need for a given amount of fabric.

Suggestions

Following is a list of some of the herbs you might want to try and the generally expected result.

PLANT	COLOR	PART USED	MORDANT
Yarrow	yellow	flowers	alum
Agrimony	yellow	leaves/ stalk	alum
Onion	orange/ green	skin of bulb	alum/ copper/ blue vitriol
Alkanet	red	roots	alum
Marigold	yellow	flowers	alum
Safflower	pink/ orange/ yellow	flowers	alum
Lily-of-the-Valley	yellow	fall leaves	chrome
Yellow Bedstraw	yellow/ pink	roots/ tops of herb	alum

Woad	blue	leaves	alum/
			potash
Broad Dock	dark yellow	roots	alum

Herb Folklore and Tips

Herbs were at one time said to be an aid to finding the perfect lifetime partner. A few leaves were stuffed under the pillow, and the dreamer, usually female, would "see" the man of her life.

You may not be seeking such a vision, but a small pillow or sachet filled with various herbs will often make the night go much easier and add a little comfort for overnight guests. Some of these comforting scents can be stuffed into a small pillow by themselves or in combination with others: Anise, Bergamot, Fennel, Lemon Verbena, Marjoram, any of the Mints, Rosemary, Lemon Thyme, Rose petals, Sage, and Chamomile.

You can also sprinkle some of the above under the lining of your dresser drawers to sweeten the clothes stored there. Also, stuff the linings of coathangers with some of the same potpourri mixture, or make your own cloth picture frames filled with your favorite scent, instead of plain cotton scraps.

Combine potpourris with different kitchen spices and hang them

Herbal potpourris and soaps sweeten all they surround.

on a long tab of gingham or burlap. Each particular scent should be placed in its own little bag and tied with a stick of Cinnamon. Hang the tab in a hallway or on the wall next to the heater for fast-acting scents.

A few handfuls of your favorite herb may be tossed into boiling water just before guests arrive to make the house a home. You might try this when trying to sell your house and the prospective buyers are on their way over. High interest rates will have nothing over you!!

A handful of potpourri in the sweeper bag will make doing the floors a pleasure.

Herbs have always been a country favorite when it comes to keeping out unwanted insects. Sprinkle a few Peppermint or Pennyroyal leaves around the house, especially near the doorways, to ward off pests. Or you can combine a few herbs to make your own insect repellent, fixing it with oil of Pennyroyal or powdered Orris root. Try a few sprigs of Tansy, Calendula, Bay, Rosemary, or Sage.

A decorative and fragrant wall wreath combines dried herbs of your choice with Rose hips and bits of clove and allspice. Arrange them with small pinecones and glue the mixture together in attractive fashion with transparent glue.

Kids who find themselves short of cash on Mother's Day may find that an herbal bath, drawn and waiting for Mom after they've served her a delectable herbal omelette in bed, will provide just the perfect touch.

Herbal bath preparations also make welcome gifts at Christmas and for birthdays or anniversaries. Make up your own labels, and don't give away all of your secrets! Sometimes the mystery is just as inviting as the bath! If you package your bath preparations, whether dry or in oil in mason jars, you can add a personal touch by making an em-

Herbal soaps make welcome gifts.

broidered cover to fit the lid. A picture of the dominant herb in the mix, or the name of the lucky recipient, will do nicely. You can lay a pressed herb under the cover to make the package smell as good outside as in.

Sachets or pillows with the name of your friend embroidered on the front make a delightful gift. Grow your own ingredients for an inexpensive gift that still requires planning and thought as well as personal consideration.

Turn a pleasant afternoon walk in the woods into a time to harvest herbs for upcoming celebrations. Always be careful to abide by the rules of foraging. And remember, herbal gifts for children should always be stuffed or filled with a selection of herbs or plants that will assuredly not harm them, should they be of the age where it's fun to try out new teeth. This especially applies to young toddlers.

Why not start a few herbs in the summer, bring them indoors, and arrange them in a lovely ceramic bowl for a Christmas gift? The plants may be culinary, aromatic, or a combination of both—sure to perk up even the most dismal spirit, as well as freshen the kitchen. The bathroom is also a good place to grow winter herbs. The occasional steam from hot water makes an excellent natural humidifier.

Planning a wedding? A few sprigs of native herbs, or your favorite "message" blossoms, will get the point across. Chicory is handsome in a nosegay, along with other fragrant herbs.

And last, but not least, a "Tussie Mussie," or hand-held flower arrangement, will help you get through almost anything with a clear head. Just keep it nearby and take a sweet sniff occasionally when circumstances mandate. In the Middle Ages it was the only way to go outside for many a noble gentleman or fine lady.

Lavender is traditionally an herbal potpourri staple.

Glossary

Alternate leaves. Those which form singly along the stem, instead of in pairs.

Annual. A plant which completes its life cycle in one season and then dies.

Anther. In a flower, the saclike portion of a stamen that holds the pollen.

Astringent. An agent that contracts body tissue.

Axil. The angle formed between the leaf and the stem.

Axillary. Growing from the axil.

Basal leaves. Those which grow at the base of the stem.

Beak. A narrow projection from some fruits or flowers.

Berry. The simple, fleshy fruit containing one or more seeds.

Biennial. A plant with a two-year life cycle.

Bract. A modified leaf at the base of a flower cluster.

Bulb. An underground stem that is swollen, mostly fleshy, with scaly leaves.

Bulbil. Little bulblike body, often forming above ground.

Calyx. A floral envelope made of sepals, usually green.

Capsule. A dry fruit with thin walls which split open, usually with one or more compartments.

Clasping. A leaf that partly or wholly surrounds the stem at its base.

Compound leaf. A leaf divided into smaller leaflets.

Corm. A modified stem, usually found underground, fleshy, thick, with scale-like leaves.

Corolla. Petals of a flower.

Creeper. A prostrate plant, that grows along the ground.

Cyme. A complex flower cluster with a round or flat top.

Decoction. An extract of an herb obtained by boiling.

Diuretic. An agent that increases the flow of urine.

Drupe. A fleshy fruit with a single stone seed.

Fruit. The ripened ovary of a plant.

Head. A dense group of flowers with short, or no stalks.

Herb. By strict definition, a nonwoody plant that dies down to the ground after each growing season. More generally, any plant that is of use to man.

Inflorescence. A flower cluster.

Infusion. The combination of an herb and water, in order to extract the properties of the herb. Water may be hot or cold.

Lanceolate leaf. One that is shaped like the tip of a lance. Long, pointed on the end, and broad in the middle.

Leaflet. One of the parts of a compound leaf.

Legume. A dry, one-celled fruit which splits open in two lines to release its seeds.

Lobed. Having indented parts, as in a leaf.

Node. Place on stem where leaves are attached.

Nut. A simple, one-seeded dry fruit with a hard shell.

Opposite leaves. Those which form in pairs along the stem.

Ovary. The part of a flower where seeds develop.

Palmate leaf. One with three or more parts, which looks like an open hand.

Perennial. A plant that lives year after year.

Petals. Basic units of the flower, flat, usually broad, and colored.

Petiole. The part of the leaf that attaches to the stem.

Pinnate leaf. One that is featherlike, or fernlike, with many leaflets.

Pistil. The female part of a flower, consisting of the ovary, style, and stigma.

Pod. A simple dry fruit that opens at maturity.

Pollen. Spores formed in the anthers of a flower.

Poultice. A compress made of the parts of an herb, applied directly to the body.

Raceme. A long flower cluster, in which the bottom flowers bloom first.

Refrigerant. Any agent which serves to cool the body.

Rhizome. A thick, fleshy underground stem which runs parallel to the surface of the ground while growing.

Root. The underground part of a plant, without nodes.

Rose hip. The fruit of the Rose bush.

Rosette. A basal cluster of leaves, produced on a very short stem.

Runner. A stem which grows on the soil surface. It can form new roots at its nodes or its tip.

Seed. The reproductive agent of a plant.

Sepal. The basic unit of the flower covering, usually green, but sometimes colored like a petal.

Shrub. A woody, low plant with many branches.

Simple leaf. One with an undivided blade.

Spathe. A large bract enclosing a flower.

Spike. An elongated flower cluster.

Stamen. The male organ of a flower, consisting of a filament and an anther. Usually many per flower.

Stigma. The tip of the pistil, where pollen collects and germinates.

Style. The narrow part of the pistil, which connects the ovary and the stigma.

Succulent. A plant with water-storing stems or leaves.

Symmetrical. In a flower, an even or circular shape.

Terminal. A flower or leaf which appears at the top of a plant.

Toothed. Having protruding, saw-toothlike edges.

Tuber. A thick, fleshy, modified stem, used by plants for food storage and propagation.

Umbel. A flat-topped flower cluster, shaped like an umbrella.

Vein. A transport tube for food, visible on a leaf.

Whorl. Three or more leaves attached at one node.

Bibliography

Andersen, Berniece A., and Holmgren, Arthur H. *Mountain Plants of North-eastern Utah*. Circular 319. Logan: Utah State University Extension Services.

Angier, Bradford. *Field Guide to Edible Wild Plants*. Harrisburg: Stackpole Books, 1974.

Angier, Bradford: *Field Guide to Medicinal Wild Plants*. Harrisburg: Stackpole Books, 1978.

Bentley, Virginia Williams. *Let Herbs Do It*. Boston: Houghton Mifflin Company, 1973.

Boxer, Arabella, and Back, Philippa. *The Herb Book*. London: Octopus Books Limited, 1980.

Clarkson, Rosetta E. *Herbs: Their Culture and Uses*. New York: The Macmillan Company, 1946.

Crockett, James Underwood; Tanner, Ogden; et al. *Herbs*. Alexandria: Time-Life Books, Inc., 1977.

Culbreth, David M. R. *A Manual of Materia Medica and Pharmacology*. Philadelphia: Lea & Febiger, 1927.

Douglas, J. D., ed. *The New Bible Dictionary*. Grand Rapids: Wm. B. Eerdmans Publishing Co., 1962.

Fox, Helen Morgenthau. *Gardening With Herbs*. New York: The MacMillan Company, 1933.

Gabriel, Ingrid. *Herb Identifier and Handbook*. New York: Sterling Publishing Co., Inc. 1975.

Gibbons, Euell. *Stalking the Healthful Herbs*. New York: David McKay Company, Inc., 1966.

Grieve, M. *A Modern Herbal*. 2 vols. New York: Dover Publications, Inc. 1971.

Harding, A. R. *Ginseng and Other Medicinal Plants*. Rev. ed. Columbus: A. R. Harding Publishing Co., 1936.

Harris, Ben Charles. *Kitchen Medicines*. New York: Pocket Books, 1968.

Hylton, William H., ed. *The Rodale Herb Book*. Emmaus: Rodal Press, Inc., 1974.

Koslaf, Leslie. "Herb and Ailment Cross Reference Chart." 4th ed. Woodmere: United Communications, 1972.

Lighthall, J. I. *The Indian Folk Medicine Guide*. New York: Popular Libary.

Lust, John B. *The Herb Book*. New York: Bantam Books, Inc., 1980.

Medsger, Oliver Perry. *Edible Wild Plants*. New York: The MacMillan Company, 1939.

Meyer, Joseph E. *The Herbalist*. Glenwood: Meyerbrooks, 1979.

Miller, Amy Bess. *Shaker Herbs: A History and a Compendium*. New York: Clarkson N. Potter, Inc., 1976.

Niering, William A., and Olmstead, Nancy C. *The Audubon Society Field Guide to North American Wildflowers*. New York: Alfred A. Knopf, Inc., 1979.

Pammel, L. H. *A Manual of Poisonous Plants*. Cedar Rapids: The Torch Press, 1911.

Porter, C. L. *Taxonomy of Flowering Plants*. 2nd ed. San Francisco: W. H. Freeman and Company, 1967.

Ralston, Nancy C., and Jordan, Marynor. *Zucchini Cookbook*. Charlotte: Garden Way Associates, 1977.

Reader's Digest Assn. *The Secrets of Better Cooking*. Canada: Reader's Digest Assn., 1973.

Rohde, Eleanour Sinclair. *A Garden of Herbs*. Boston and New York: Hale, Cushman & Flint, 1936.

Scully, Virginia. *A Treasury of American Indian Herbs*. New York: Crown Publishers, Inc., 1970.

Weslager, C. A. *Magic Medicines of the Indians*. New York: The New American Library, Inc., 1973.

Wieand, Paul. R. *Folk Medicine Plants Used in the Pennsylvania Dutch Country*. Allentown: Wieand's Pennsylvania Dutch, 1961.

Wood, Horatio C., and LaWall, Charles H. *The Dispensatory of the United States of America*. 21st ed. Philadephia and London: J. B. Lippincott Company, 1926.

Appendix

Here's a list of the herbs and plants found within the text, arranged alphabetically by common names. This list should be of more value to those who are unfamiliar with the scientific classifications. The corresponding genus-species names are included, however, along with the page of location in the text.

Sweet Basil	*Ocimum basilicum*	142
Sweet-Briar Rose	*Rosa eglanteria*	222
Sweet Cicely	*Osmorhiza claytoni*	262
Sweet Goldenrod	*Solidago odora*	71
Sweet Marjoram	*Origanum majorana*	144
Sweet Weed	*Althaea officinalis*	189
Sweet Woodruff	*Asperula odorata*	226
Tarragon	*Artemisia drancunculus*	58
Teaberry	*Gaultheria procumbens*	119
Teasel	*Dipsacus sylvestris*	117
Thorny Pigweed	*Amaranthus spinosus*	37
Thunder Plant	*Sempervivum tectorum*	115
Top Onion	*Allium cepa, var. aggregatum*	173
Trout Lily	*Erythronium americanum*	180
True Chamomile	*Anthemis nobilis*	52
True Lavender	*Lavandula officinalis*	131
True Watercress	*Nasturtium officinale*	95
Velvetleaf	*Abutilon theophrasti*	188
Virginia Rose	*Rosa virginiana*	221
White Clover	*Trifolium repens*	168
White Dog-Tooth Violet	*Erythronium albidum*	180
White Goosefoot	*Chenopodium album*	110
White Mustard	*Brassica alba*	87
White Sweet Clover	*Melilotus alba*	164
Wild Basil	*Cunila origanoides*	127
Wild Bergamot	*Monarda fistulosa*	140
Wild Carrot	*Daucus carota*	258
Wild Dog Rose	*Rosa canina*	222
Wild Garlic	*Allium canadense*	171
Wild Geranium	*Geranium maculatum*	121
Wild Ginger	*Asarum canadense*	47
Wild Leek	*Allium tricoccum*	177
Wild Marjoram	*Origanum vulgare*	145
Wild Sage	*Salvia lyrata*	150
Wild Sarsaparilla	*Aralia nudicaulis*	43
Wild Senna	*Cassia marilandica*	162
Wild Thyme	*Thymus serpyllum*	155
Wintergreen	*Gaultheria procumbens*	119
Winter Savory	*Satureia montana*	154
Woad	*Isatis tinctoria*	92
Wolfsbane	*Aconitum uncinatum*	211
Wood Betony	*Betonica officinalis*	126
Wood Sage	*Teucrium canadense*	157
Wood Strawberry	*Fragaria vesca*	217
Woody Nightshade	*Solanum dulcamara*	245

Index